# Quality Lighting for
# High Performance Buildings

# Quality Lighting for High Performance Buildings

Michael Stiller

THE FAIRMONT PRESS, INC.

CRC Press
Taylor & Francis Group

**Library of Congress Cataloging-in-Publication Data**

Stiller, Michael, 1961-
Quality lighting for high performance buildings / Michael Stiller. -- 1st ed.
p. cm.
Includes bibliographical references and index.
ISBN 0-88173-665-1 (alk. paper)
ISBN 0-88173-666-X (electronic)
ISBN 978-1-4665-0130-0 (taylor & francis distribution : alk. paper)

1. Dwellings--Lighting. 2. Commercial buildings--Lighting. 3. Electric lighting--Energy conservation. 4. Light in architecture. I. Title.

TH7975.D8S75 2012
621.32′2--dc23

2011046474

*Quality Lighting for High Performance Buildings /Michael Stiller.*


Published by The Fairmont Press, Inc.
700 Indian Trail
Lilburn, GA 30047
tel: 770-925-9388; fax: 770-381-9865
http://www.fairmontpress.com

Distributed by Taylor & Francis Ltd.
6000 Broken Sound Parkway NW, Suite 300
Boca Raton, FL 33487, USA
E-mail: orders@crcpress.com

Distributed by Taylor & Francis Ltd.
23-25 Blades Court
Deodar Road
London SW15 2NU, UK
E-mail: uk.tandf@thomsonpublishingservices.co.uk

Printed in the United States of America
10 9 8 7 6 5 4 3 2 1

ISBN 0-88173-665-1 (The Fairmont Press, Inc.)
ISBN 978-1-4665-0130-0 (Taylor & Francis Ltd.)

# Table of Contents

Acknowledgments .......... ix
Introduction .......... xi

**PART I—QUALITY LIGHTING** .......... 1

**Chapter 1—What is Lighting Design** .......... 3

**Chapter 2—Understanding Light (Luminance, Illuminance, & Lumens)** .......... 11
Lumens & Candelas .......... 11
Illuminance & Luminance .......... 13

**Chapter 3—Glare & Contrast** .......... 15
If there's so much light, why can't I see a thing? .......... 15
It's All Relative .......... 15
The Glare Zone .......... 16
Luminous Intensity .......... 18
The Luminance Factor .......... 26
Minimizing Glare— It's not just about the lighting. .......... 28

**Chapter 4—Visual Comfort & Visual Interest** .......... 29
What is Visual Comfort & Visual Interest? .......... 29
The Obsession with Horizontal Footcandles .......... 30
Visual Comfort & Visual Interest: How do we get there? .......... 32
Luminance Ratios .......... 35
Luminance Ratios & Daylight .......... 38

**Chapter 5—Color & Light** .......... 41
Color Temperature .......... 41
CRI & Color Quality .......... 43

**PART II—HOW MUCH LIGHT DO WE NEED & WHERE DO WE NEED IT?** .......... 45

**Chapter 6—Lighting + Space: Calculating the Results** .......... 47
Avoiding the tendency to over-light .......... 47

The results of our lighting depends on more than our lighting.....47
Choosing a Calculation Method.....50
Inverse Square Law.....51
Zonal Cavity Method.....52
Computer-Based Calculations.....56
IES Files.....58
Radiosity & Ray Tracing.....59

**Chapter 7—Target Illuminance Levels**.....65
How much is enough?.....65
A Moving Target.....66
Recommendations for Today.....69

**Chapter 8—Task Lighting**.....73
Getting the light where it's needed.....73
Localized General Lighting.....75
Furniture-mounted Task-Ambient Systems.....78

**PART III—SUSTAINABILITY & ELECTRIC LIGHTING SOURCES**.....81

**Chapter 9—Choosing Lamp Types & Sources**.....83
Design Factors.....83
Luminous Efficacy.....84
Color Quality.....84
Life Cycle Assessment (LCA).....86

**Chapter 10—Lamps, Source Types & Relative Photometry**.....89
Incandescent Lamps.....89
Halogen Lamps.....91
Metal Halide Lamps.....93
Sodium Vapor Lamps.....95
Fluorescent Lamps.....95
Ballast Factor & Ballast Efficiency Factor.....97
Formats & Efficacies.....100
High Performance T8 Systems.....101
Self-ballasted Compact Fluorescent Lamps.....102
Magnetic Induction Lamps.....102

Light Emitting Diodes....106
Sources, Applications & Fixture Efficiency....107
Relative Photometry....108

**Chapter 11—LED (SSL) Lighting**....111
What is an LED?....111
Color Mixing & White Light....112
Applications for RGB LEDs....113
High Brightness White LEDs....114
LED Light Engines....115
Color Consistency....116
Thermal Management....117
LED Performance & Lifetime (L70)....118
Absolute Photometry &
The Luminous Efficacy of LED Luminaires....120
New Photometry, New Standards....122
LCA & Modularity....124
LED Replacement Lamps....127
Lumen Distribution & LEDs
(a natural fit for a sustainable design strategy)....131

**PART IV—SUSTAINABLE APPLICATIONS:
DAYLIGHTING & LIGHTING CONTROLS**....133

**Chapter 12—Daylighting**....135
What is Daylighting?....135
Effective Illumination....136
Architectural Daylighting Treatments....139

**Chapter 13—Lighting Controls**....145
The Case for Enhanced Lighting controls....145
Daylight Harvesting....149
Occupancy Sensors & Vacancy Sensors....158
Fluorescent Lamps and Occupancy Sensors....162
Time-of-day Controls....163
Load Shedding....165
Continuous Dimming & Step Dimming....167
Lighting Controls, Ballasts, Drivers & Sources....170

Low-voltage Control Systems & Protocols .................. 174
Analog Controls .................. 174
Digital Distributed Controls .................. 178
Wireless Controls .................. 183
Dimming Ballast and Driver Performance .................. 186

**PART V—BUILDING GREEN** .................. 189

**Chapter 14—Model Codes,**
**Code-Language Standards & Energy Codes** .................. 191
Model Codes .................. 191
ANSI/ASHRAE/IESNA 90 .................. 194
COMcheck & REScheck .................. 200
California's Title 24 .................. 201

**Chapter 15—Paths to High Performance Green Buildings:**
**Advocacy Groups, Advanced Energy Design Guides, Green**
**Construction Codes & Green Building Rating Systems** .................. 203
The 2030 Challenge & Zero Net Energy Buildings .................. 203
ASHRAE Design Guides & Energy Star .................. 205
Green Building Rating Systems .................. 207
Green Building Standards
& Green Model Construction Codes .................. 212

**Appendix A—Color Illustrations** .................. 215

**Appendix B—Lighting Calculations & Calculation Software** .................. 221

**Appendix C—Resources** .................. 230

**Index** .................. 237

# Acknowledgements

Thanks go to Jennifer Stiller for all her support and for her detailed line edits to this manuscript. I would also like to thank the following people and firms for their assistance in the production of this book, either as readers, professional advisors, or by securing permissions for the use of some of the material herein. They are, in no particular order: Jim Toole, Dave Speer, Amanda Beebe, Nancy Bea Miller, Hayden McKay, Summer Marshall, Clayton Gordon, L. Michael Roberts, Edgar Chan, Jennifer Psycher, John Moon, Joseph Herrin, Carrie Martinelli, Michelle Murray, Mark Silvey, Bill Welch, Heliotrope Architects, Cree Inc., BetaLED, Koncept Technologies, Corelite/Cooper Lighting, Wattstopper, Leviton Manufacturing Inc., Illuminating Engineering Society of North America, Association of Energy Engineers, The Lighting Quotient, and Lighting Analysts, Inc.

Special thanks go to Stan Walerczyk, Dawn De Grazio, and John Newman for their indispensible contributions to the sections on electric lighting sources, lighting calculations, and quality lighting.

# Introduction

According to the U.S. Department of Energy's *Buildings Energy Data Book*, in the United States the buildings sector accounted for over 40% of primary energy consumption in 2010. Lighting in our homes accounted for 10% of this usage, and in our commercial buildings it accounted for 17.4%. These are significant numbers, and there can be no doubt that reducing our energy use will carry with it a myriad of benefits, be they economic (saving us money), environmental (reducing our carbon footprint), or political (reducing our dependence on foreign oil). There are many strategies we can employ to this end, but the real challenge comes in reducing our usage while simultaneously retaining, and even improving, our quality of life.

*Quality Lighting for High Performance Buildings* is an introduction and guide to the considerations and principles all lighting designers, architects, and engineers should apply to their projects when specifying a lighting system that is meant to be energy-efficient, sustainable, visually comfortable, and generally pleasing to a building's occupants. This book is written in non-technical language wherever possible, so that it may be read by building design professionals, students, and interested members of the general public alike. The practice of lighting design is a complex craft, as much art as it is science, and in that sense it cannot be taught solely through books, or in the classroom. But it is my belief that any of us can begin to participate in this most engaging of pursuits with a little knowledge and an observant eye.

Starting with a primer on the basics of lighting design, this book provides an introduction to the properties of different lighting sources (including LED lighting), lighting fixture efficiencies, relative and absolute photometry, daylighting, lighting controls, and energy codes.

High performance buildings are energy-efficient, have low short-term and long-term life cycle costs, are designed and situated in such a way as to minimize disturbance to their natural surroundings, and are constructed from and operated with materials whose manufacture and disposal have a low impact on the natural environment. In addition to being the product of a sustainable design process, high performance buildings must also provide a healthy and productive indoor environment for their occupants. Quality lighting for high performance buildings is lighting designed with all of these principles in mind, and it must take into account economics, energy efficiency, sustainability, and the needs of those who live and work within our built environment.

# Part I

# Quality Lighting

# Chapter 1

# What is Lighting Design?

## A DEFINITION

*Lighting design is the specification of a system of luminaires and controls to create illumination appropriate to a given environment.* This means that lighting designers choose which lighting fixtures should go where (specification of a system of luminaires), as well as how they are grouped and which should be on at a given time and at what levels of intensity (controls). But how does the designer determine what illumination is *"appropriate to a given environment"?* Historically, architectural lighting designers, and to a great degree electrical engineers performing in this capacity, have been concerned with providing enough illumination for a specific visual task. Quantity was the key. And the question was, simply: how much light is enough? It's been a long time since we've considered lighting design in such simple terms. As a culture we accept that good lighting is important in many other ways, whether it's to create a comfortable, productive environment, or to set a mood. Many other factors are central to the design of a quality lighting system: color temperature, accurate color rendering, volumetric quality, and contrast ratios, to name a few. Yet even so, many of those outside of the profession—and some within it—still focus on the quantity rather than the quality of light delivered by a given system. And it's easy to see why. Lighting is ethereal. It has physical properties but we can't touch it. It helps us see, but outside the context of the world of objects, which reflect light back to us to create an image of those objects in our minds, light is, by itself, invisible.

## A CASE IN POINT

Recently, at a visit to a medical office comprised of a very well designed suite of examination rooms, I had a brief conversation with one of the staff members about the lighting. This is a facility which clearly has benefited from an integrated design process, with a lot of indirect lighting positioned so as to illuminate appropriately reflective surfaces and light the space with pleasing contrast ratios and as little direct light and glare as possible. The designer specified cove treatments, wall washers, recessed vertical fluorescent lighting in the hallway walls with acrylic diffusers, and backlit mirrors in the restrooms. The examination rooms each contained a fluorescent fixture with diffusing lenses and shielding louvers over the exam area, where higher illuminance levels are necessary. The consultation areas were bathed in an ambient light that helped the doctors to accurately see their patient's skin tones and facial expressions. The staff's work-stations were outfitted with under-shelf task lighting. There were decorative pendants that simultaneously provided a gentle accent as well as task lighting on the public-facing counters. The ambient illuminance levels were low—between ten and fifteen footcandles in the circulation areas—yet there was plenty of light for patients of all ages to negotiate their way around. When I asked the staff member whether she liked the lighting in these new offices, and if she thought it was good, she readily agreed that it was. But the way she described the complex, layered, low-ambient lighting design was to simply say it felt "brighter" than her old office—which it undoubtedly wasn't. What she clearly meant was that she could see better. Maybe she was aware that the lighting in her present facility went a long way towards setting a mood. This was a modern, clinical research facility in New York City. The design was one that lent a serious and reassuring air to a place where visitors bring their greatest anxieties and concerns for their own well-being—a doctor's office. It was clear from our conversation that the staff member had an awareness of these factors. She knew good lighting when she saw it, but thinking about lighting and articulating her thoughts in terms other than "brightness" was something she had difficulty doing.

Such is the case for many of us in the design professions. Most architects, interior designers, and engineers are aware that a well-designed lighting system is important for many reasons, and that while having enough light is necessary for us to be able to see, lighting can also greatly affect the way we experience our environment in numerous other ways. Anyone who has walked into a poorly maintained store where the fluorescent lighting is a hodgepodge of different color temperatures, or where the lamps have not been changed in many years, and after burning way past their expected lifespan emit only a fraction of their original light output, will tell you that their experience of that store was diminished as a result of the lighting quality. We all know that poor lighting can make a customer feel uncomfortable buying a product that might require advanced support from a store owner; or prompt a client to hire one business over another for a project; or tip the balance for a tenant deciding whether or not to lease a particular apartment or office. But for most of us, the only way we are able to discuss lighting is in terms of whether

**Figure 1-1. Too much light from a flashbulb creates direct glare and obscures the faces of a crowd at a sporting event, Photo credit: Arcimboldo (Own work) [CC-BY-3.0 (www.creativecommons.org/licenses/by/3.0)], via Wikimedia Commons**

or not there is enough of it, when in fact the quality of the lighting is as important as the quantity of lighting.

The fact is, in certain situations too much light, coming from the wrong direction, can make it difficult to see. Glare from a direct line of sight to a light source, or lighting that produces high contrast ratios within an environment, can inhibit our ability to discern objects or features that fall within the shadows. Anyone who has had the experience of talking with someone who is standing with their back to a window on a brightly lit day will recognize this right away. Though there may be plenty of light, we still struggle to see those objects that are right in front of us. But though this is a common phenomenon, and one we take for granted in our everyday experience, understanding the nuances of light, and how we experience the lit environment in terms other than brightness, remains a challenge for most people.

Now, let's circle back to our definition of lighting design: *"The specification of a system of luminaires and controls to create*

**Figure 1-2. Light pouring in a window obscures the face of a painter at work, Photo Credit: Hugh Downs, Nancy Bea Miller at Window, 2009**

*illumination appropriate to a given environment."* What constitutes appropriate illumination? Clearly having the right amount of light is important. We need a certain amount of light to see, and, as we will discuss later in the book, more light to see fine details, and even more light as we get older. We may also need different amounts of light depending on the mood we want to convey. Certainly a romantic restaurant will be lit a good deal less brightly than an automobile showroom. Most of us can recall visiting both brightly and dimly lit environments where our visual experience was good, and those where it was not. So if it's not just about the quantity, what then constitutes quality lighting?

## WHAT IS QUALITY LIGHTING?

Quality lighting is the result of an integrative design process in which the building design, interior design, and lighting design evolve together to create a built environment that is sustainable, productive, and healthy for the people who occupy it. Quality lighting must support the human activities that each environment is designed for. As designers we start with the assumption that lighting is for people, not just buildings, and so quality lighting must satisfy the needs of the occupants even as it provides aesthetic appeal and reveals and enhances the building's architectural form.

## WHAT IS SUSTAINABLE DESIGN?

A building design is sustainable when it results in lowered energy use and lessened environmental damage during the construction phase and operational lifetime of the building. The concept of sustainability is based on an understanding that there are finite resources, and space, at our disposal, and that we need to limit our consumption of these resources to the amount that can

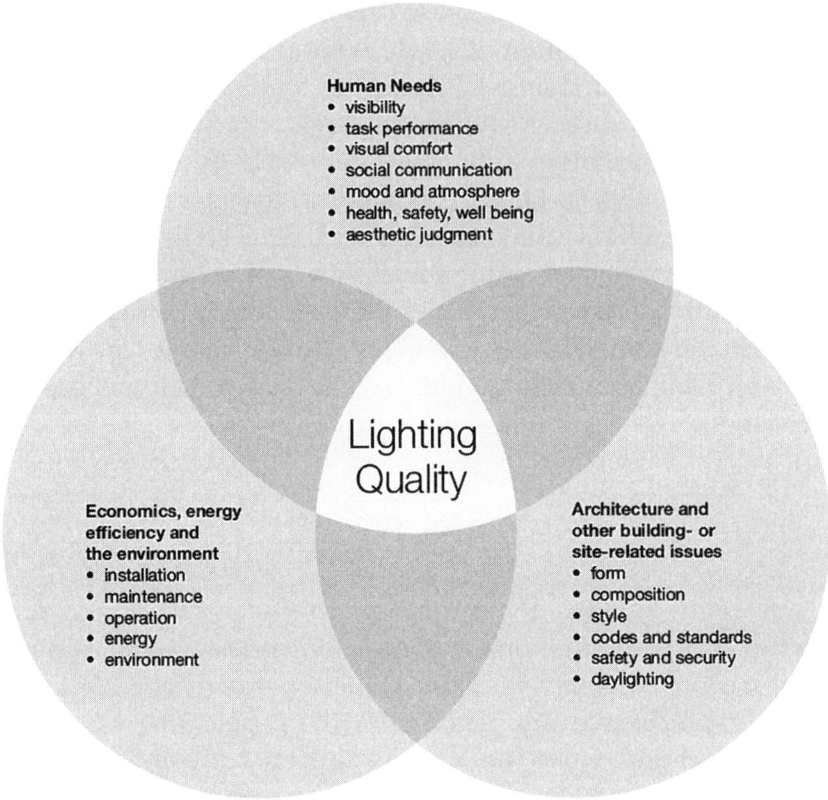


**Figure 1-3. From the IESNA Lighting Handbook, 9th Edition by permission from the Illuminating Engineering Society of North America**

be regenerated, both for our health and to insure we do not run out. This applies to energy, potable water, air, and arable land. To be sustainable, a quality lighting design must be energy-efficient. But using high efficiency lighting and avoiding wasted energy is not enough. The lighting equipment should also be manufactured in a sustainable way that doesn't waste resources or emit pollutants into the water or atmosphere. It should include as few toxic materials as possible, generate as little waste as possible, and it should be sourced as close to the construction site as possible to avoid expending unnecessary energy in transport.

Good design is key to keeping sustainable lighting sustainable. If lighting is for people, and not just buildings, then it follows that if it does not satisfy their human requirements a building's occupants may modify the lighting in ways that are often contrary to the designer's intent, and in doing so they may negate any energy efficiencies that would otherwise be realized. These modifications can, and often do, take the form of additional electric lighting that may be from inefficient sources, or disabled energy efficiency controls. An incorrectly commissioned occupancy sensor that automatically turns the lights off in a room while the occupants are still present is a prime example. This will not have to happen more than once or twice before the people who regularly use this room find a way to disable the occupancy sensor, negating any benefit that control device may have brought to the design in the first place. A poorly designed lighting layout

**Figure 1-4. An office worker who has taken the task-lighting of his workspace into his own hands. Photo Credit: Hayden McKay**

that wastefully directs too much light to an open office's circulation areas, and not enough to the employee's desks, may result in the facility manager or the employees themselves developing ad-hoc task lighting solutions that are not energy-efficient, and that may push the office out of compliance with the local energy code. And a lighting design that only takes into account the work-plane lighting levels, with no consideration for the contrast ratios between the task area and surrounding surfaces, may have the negative result of creating an uncomfortable environment that leads to employee fatigue and reduced productivity.

So what are the elements of a quality lighting design? As we will discuss in this book, in addition to being energy-efficient and providing the right amount of light, a quality lighting design must minimize glare and manage contrast to create a visually comfortable environment. It must also take into account the color temperature or color quality of the lighting sources employed so that colors are rendered accurately, and the occupants, and the environment itself, look good. It must provide the right amount of light to set the correct tone and to support the visual tasks that will be performed. It should make effective use of any available natural daylight. And it should incorporate a system of controls that allows the lighting to be dimmed, or switched off, as required to suit individual preferences and minimize energy use as much as possible throughout the day.

# Chapter 2

# Understanding Light:
## (Luminance, Illuminance and Lumens)

LUMENS & CANDELAS

It's important to clarify some of the terms we will use throughout this book, and in doing so develop a clearer understanding of how electric lighting works. First, let's start with the light source and luminaire. The light source, or lamp (commonly called the light bulb), produces a certain amount of visible radiant energy, or light, which we measure in *lumens*. The lamp, in combination with the luminaire (a fancy term for a lighting fixture), will direct and define how these lumens are projected. Some lamps and luminaires will project the lumens in an omnidirectional pattern (in all directions), like a household light bulb in a table lamp with a translucent lampshade; and some will project them as a narrow beam of light or a well-defined spot, like a reflector lamp in a directional track light. If the number of lumens is the same in both of these cases, the light striking the nearby surfaces (provided they are of an equally reflective color and equally far from the light source) will appear brighter in the case of the reflector lamp and track light, and the illuminance will be greater, as the same amount of light, or lumens, will essentially be concentrated in a smaller area.

A further discussion of brightness, luminance, and illuminance will follow, but first let's talk a little more about the light coming from our luminaire and source. The overall light coming from our luminaire, the lumens, can be further defined as *candelas,* or candlepower. A candela is really a measure of the concentration of lumens in a particular direction. To use our previous examples of a table lamp and a directional track light: though the number of lumens might be the same, when they are concen-

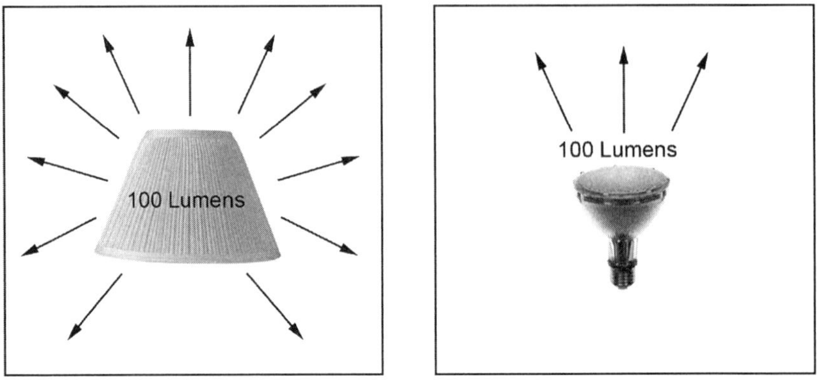


The Same amount of light spread over a greater area = less light per unit of area

The Same amount of light spread over a smaller area = more light per unit of area

**Figure 2-1. As the projection of light, or lumens are spread over a greater area, the illuminance will decrease, as each unit of area of the surrounding surfaces receives a lesser concentration of the overall light that is emitted.**

trated into a narrower beam by the track light, the candlepower, or the number of candelas, will be greater in the direction of that beam than when they are projected in an omnidirectional pattern, or a greater number of directions, around the room. Candlepower is a constant measurement, and does not change with distance. Even as the light will spread over distance, the number of lumens, and therefore candelas, coming from the source in that particular direction will remain the same.

Once the light has left our luminaire, it spreads over a larger area as it travels through space. We have all had the experience of shining a flashlight on the ground in front of us, and then shining it on an object some distance away; the beam of light widens out and it appears less bright. Though the number of lumens in the beam of light are constant, as the beam spreads over a greater area the same lumens are also spread over a greater area and fewer will fall in any one place. The result is a dimmer appearance of light. In addition, certain objects will appear more or less bright depending on their color and their other qualities of reflectance. These are examples of both illuminance and luminance.

## ILLUMINANCE & LUMINANCE

Illuminance is the amount of light per unit area that is projected *onto* a given surface, while luminance is the amount of light per unit area that is emitted or reflected back to the viewer *from* a given surface. Illuminance has both a metric and imperial expression. The metric expression, lux, is the measure of the amount of light, or number of lumens, that falls onto a surface the size of a square meter; and the imperial expression, footcandles, is the amount of light that falls onto a surface the size of a square foot. Both the illuminance on a surface and the luminance of a surface will vary as the distance is varied between the surface and the source of light, as they are both dependent on the concentration of light, or lumens, that falls that surface. As we have demonstrated with our flashlight, the number of lumens falling on a given area will decrease as we move our light source farther from the object being lit, and the beam, made up of a constant number of lumens, spreads out to cover a greater and greater area. The luminance of a surface will also be affected by other factors, namely that surface's translucence (in the case of back-lit materials), reflectance (or color) and specularity (or glossiness).

*Luminance* is measured on a metric scale: candelas per square meter ($cd/m^2$), and is the amount of light that is emitted or reflected from a square meter of a given surface in a given direction. Luminance is also sometimes defined as the perceived brightness of an object. Brightness is not a precise technical term, but it's one we use every day. What many people may not realize is that our experience of brightness does not necessarily relate in a direct way to the amount of light in a particular environment. A fully lit room with dark walls and a dark ceiling may not feel very bright though there is plenty of light (or a high level of illuminance) on all of the room's surfaces. It's just that most of it is being absorbed by the dark surrounding surfaces, which have a low luminance. As we've said, what we call brightness, or luminance, is really the amount of light that is reflected, or emitted, towards our eyes from the objects and surfaces surrounding us, while illuminance is a truer measure of the amount of light in the environment to begin with. When we talk about the amount of light necessary to see certain details, or accomplish certain tasks, we will use the terms *lux* and *footcandles,* which are measures of illuminance. When we talk about contrast and glare, or the perceived brightness of a surface or object, luminance, measured in candelas per square meter *(cd/m²),* is the term we will most often use.

To return to our example of the room with a dark ceiling and dark walls: the light, or lumens, generated by the interior lighting (or let into the room through a window) are projected in a defined pattern made up of concentrations of candelas. This light falls onto a given area of the wall's surface as illuminance, and is reflected back to the viewer as perceived brightness, or luminance. Except in the case of a self-luminous surface, the amount of illuminance striking, or passing through a surface will have a direct relation to its luminance, but the reflective (or transmissive) properties of that surface are big factors in determining the degree of this luminance as well. This is why I like to say that light is itself invisible. We can have the brightest light source possible, but if the light generated is not reflected, or emitted, back to us, as luminance, there is nothing for us to see.

# Chapter 3

# Glare & Contrast

## IF THERE'S SO MUCH LIGHT, WHY CAN'T I SEE A THING?

Glare and contrast, two of the biggest issues we deal with as lighting designers, are directly related to one another. Glare occurs when an object in our field of view is overly bright in relation to its surrounding environment. This creates visual discomfort and reduces our ability to see the other objects around it. Common causes of glare are an unshielded light source, like a bare light bulb (or lamp), direct sun (or even diffused daylight) streaming into an interior environment, poorly designed street or parking lot lighting, or car headlights at night.

**Figure 3-1. Direct glare from a poorly designed lampshade that does not fully cover the light source.**

## IT'S ALL RELATIVE

An important consideration to keep in mind is that glare is relative. Many of us have experienced being temporarily blinded by the headlights from oncoming traffic when traveling at night on a narrow country road. But when traveling along that same country road in the middle of a brightly lit day, those headlights will be hardly noticeable. That is because while we have a great ability to see under a broad range of lighting levels, our sight is limited to a relatively small portion of that range at any one time. Most of us can see well enough to get around with as little as one footcandle of illuminance on the walls and floor of a given area, though older people and those with visual impairments will need more. And we can see perfectly well outdoors on a brightly lit day in which the illuminance levels might be as high as, or even higher than, 10,000 footcandles. That's a ten thousand-fold difference! But though we can see pretty well under either of these conditions, we can't see under both of them at the same time. As anyone who has walked from a sunny outdoor environment into a darkened room knows, our eyes need a minute or so to adjust before we can see again. And walking back outside can be almost painful, as the sun will now create a blinding glare.

So then, glare is not necessarily caused by objects of a specific brightness but is actually the result of a very high contrast ratio between a specific object or surface that is much brighter than the rest of the objects or surfaces around it. We can break the phenomenon of glare down into two main categories: direct glare, and reflected glare. Direct glare is caused by a direct line of sight to very brightly illuminated objects, like lighting fixtures or a brightly lit outdoor environment seen through a window; and reflected glare is caused by bright reflections on specular or glossy surfaces, like computer screens or marble floors.

## THE GLARE ZONE

One of the main things we can do to avoid direct glare when designing a lighting system is to specify fixtures with appropriate shielding or diffusers that block any direct line of sight to the lamps, reflectors, and other especially bright components of the fixture. With downlights, the traditional rule of thumb has been to limit a direct line of sight to lamps and other high-brightness

Figure 3-2. Reflected glare from a marble floor, Photo Credit: Summer Marshall

Figure 3-3. Reflected Glare on a computer display at a German rail station makes it impossible to read the arrival and departure information displayed, Photo credit: Panjasan (Own work) [CC-BY-SA-3.0 (www.creativecommons.org/licenses/by-sa/3.0)], via Wikimedia Commons

fixture components to the area between 0 and 55 degrees, as measured from vertical. Above 55 degrees, these bright areas can create glare, as they are more likely to be in our field of view while we go about our tasks and travel through the space. With direct/indirect lighting, i.e. lighting fixtures that simultaneously throw light up and down to illuminate the ceiling as well as the floor and horizontal work-plane, we also try to limit any direct line of sight to lamps and high-brightness fixture components to the area above 90 degrees. Figures 3-4 and 3-5 illustrate both of these cases. By limiting the observer's line of sight to bright lighting fixture components to the area between 0 and 55 degrees we create a scenario whereby the potentially glare-producing parts of the fixture are above our heads, and out of our field of view, when they are visible at all.

## LUMINOUS INTENSITY

The amount of light coming from a lighting fixture at specific angles, expressed as a certain number of candelas, is called luminous intensity. We can use luminous intensity charts, sometimes called candela or candlepower distribution charts, and often included as part of a lighting fixture's specification sheet, to determine if a particular fixture is likely to cause glare in a particular situation. These charts will tell us how much light is projected from the fixtures along each angle, including the area we defined as the glare zone: the angles between 55 and 90 degrees. But what is the maximum luminous intensity that we should accept from a downlight in the glare zone? As we've already shown, glare is relative. And so the answer is, it depends. For example, in 2004 the Illuminating Engineering Society of North America (IESNA) determined that in open offices where much of the work was done on computers using VDTs (video display terminals), the luminous intensity of lighting fixtures should not exceed a maximum of 300 candelas at 55 degrees.[1] This was a way to minimize the reflected glare on the VDTs from overhead lighting.

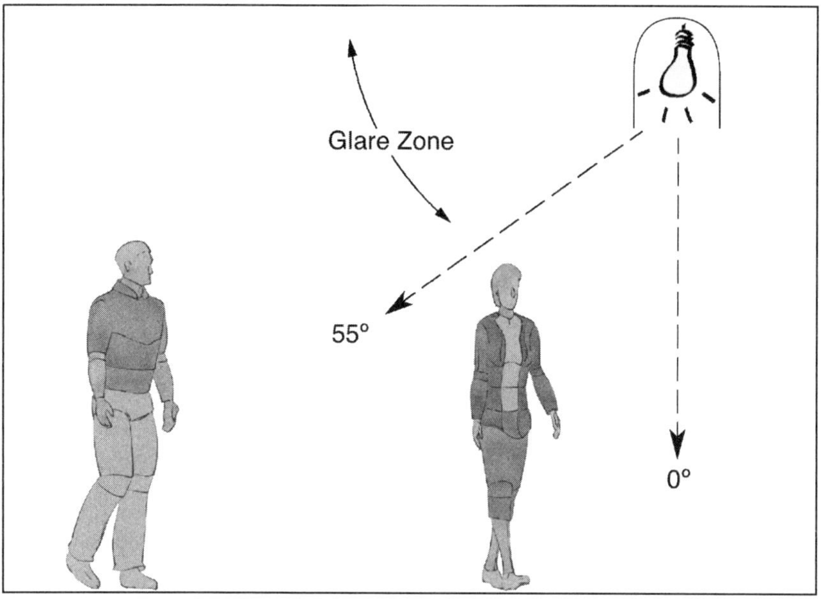


Figure 3-4. Glare Zone for a downlight.

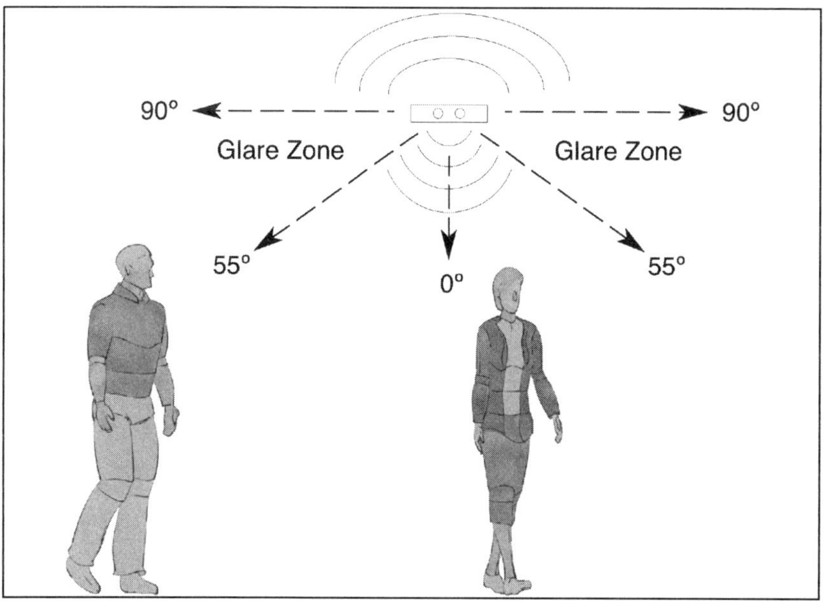


Figure 3-5. Glare Zone for a direct/indirect pendant fixture.

(Video display terminal is a technical term for computer monitors, the older versions of which featured glossy, curved screens which tended to reflect a great deal of the ceiling and the lighting fixtures back to the viewer. These older VDTs are also called CRTs, which stands for cathode ray tube and is the name of the technology that creates the image in these older monitors.) Figures 3-6 and 3-7 show luminous intensity charts for a downlight and a direct/indirect pendant fixture that meet this criterion, and can be compared with the illustrations of these treatments in Figures 3-4 and 3-5.

Outside of open offices with CRTs, or glossy VDTs, we often limit the luminous intensity of lighting fixtures to 600 candelas at 65 degrees, and in high ceilinged spaces, like warehouses, we try

Candlepower Distribution

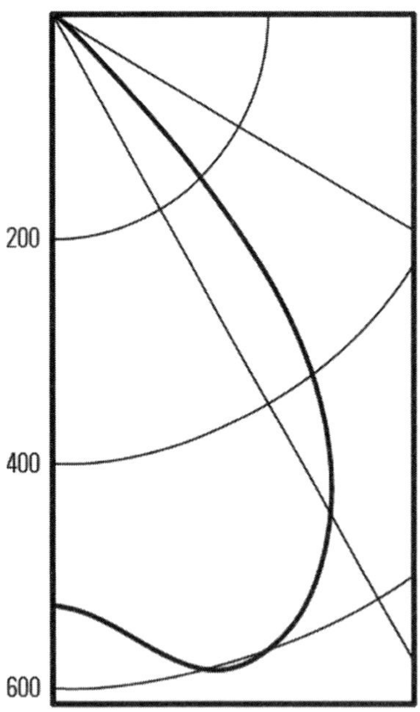


| ANGLE | MEAN CP | LUMENS |
|---|---|---|
| 0 | 526 | |
| 5 | 542 | 54 |
| 10 | 578 | |
| 15 | 606 | 170 |
| 20 | 599 | |
| 25 | 568 | 259 |
| 30 | 515 | |
| 35 | 439 | 269 |
| 40 | 329 | |
| 45 | 154 | 123 |
| 50 | 30 | |
| 55 | 12 | 12 |
| 60 | 6 | |
| 65 | 3 | 4 |
| 70 | 2 | |
| 75 | 1 | 1 |
| 80 | 0 | |
| 85 | 0 | 0 |
| 90 | 0 | |

**Figure 3-6. Luminous Intensity Chart for a recessed downlight, as illustrated in Figure 3-4. The chart shows just how much light is projected from the lighting fixture along the various angles between 0 and 90 degrees.**

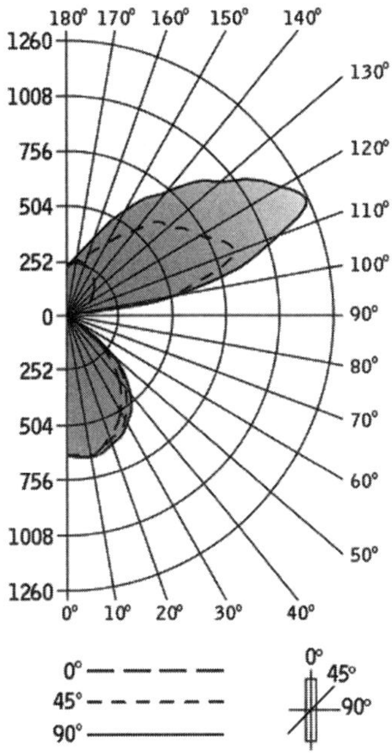


| Vertical Angle | 0° | Horizontal Angle 22.5° | 45° | 67.5° | 90° |
|---|---|---|---|---|---|
| 0° | 642 | 642 | 642 | 642 | 642 |
| 5° | 647 | 647 | 647 | 648 | 647 |
| 15° | 621 | 625 | 634 | 643 | 644 |
| 25° | 556 | 560 | 585 | 601 | 603 |
| 35° | 441 | 465 | 496 | 515 | 524 |
| 45° | 284 | 303 | 340 | 363 | 369 |
| 55° | 105 | 111 | 124 | 128 | 128 |
| 65° | 37 | 43 | 45 | 48 | 48 |
| 75° | 10 | 14 | 17 | 18 | 18 |
| 85° | 0 | 1 | 2 | 5 | 5 |
| 90° | 0 | 0 | 4 | 7 | 7 |
| 95° | 16 | 159 | 132 | 112 | 109 |
| 105° | 55 | 381 | 689 | 847 | 863 |
| 115° | 103 | 375 | 806 | 1159 | 1259 |
| 125° | 143 | 353 | 688 | 975 | 1097 |
| 135° | 184 | 303 | 614 | 793 | 880 |
| 145° | 216 | 286 | 445 | 601 | 663 |
| 155° | 243 | 270 | 346 | 426 | 464 |
| 165° | 253 | 260 | 279 | 307 | 324 |
| 175° | 248 | 248 | 248 | 248 | 248 |
| 180° | 233 | 233 | 233 | 233 | 233 |

**Figure 3-7. Luminous Intensity Chart for a direct/indirect pendant fixture, as illustrated in Figure 3-5. The chart shows just how much light is projected from the lighting fixture along the various angles between 0 and 180 degrees.**

to limit the luminous intensity to 1,000 candelas at 65 degrees. But just as glare is relative, these are relative targets. Figure 3-8 shows the luminous intensity chart for an industrial high-bay fixture that projects over 1,200 candelas at 65 degrees. Note that this fixture has a small percentage of its distribution as uplight, which will help to light the ceiling, and minimize the contrast between the ceiling and the fixture itself. Whether or not this will create a glare condition will depend on the actual height of the fixtures and the color of the ceiling and walls. If the surrounding surfaces are light colored and highly reflective (but not specular), then the brightness of the fixture will be relatively less than if the surrounding surfaces were dark. Similarly, the higher the lighting fixtures are, the farther away they will be from the viewer when he or she is in the "glare zone," diminishing the glare effect. But in a dark colored environment, or when the fixtures are lower, the likelihood that these fixtures would create a serious glare condition would be far greater.

### An Evolution of Design

As our technology changes, so do our lighting specifications. In the 1990's, with the ubiquity of CRT monitors in open offices, *parabolic troffer* lighting fixtures were widely adopted as the recessed lighting fixture of choice. The name *troffer* comes from a combination of the words *trough* and *coffer*, and the parabolic troffer is a very efficient lighting fixture designed to provide extremely good glare control. These fixtures have a recessed, open-faced

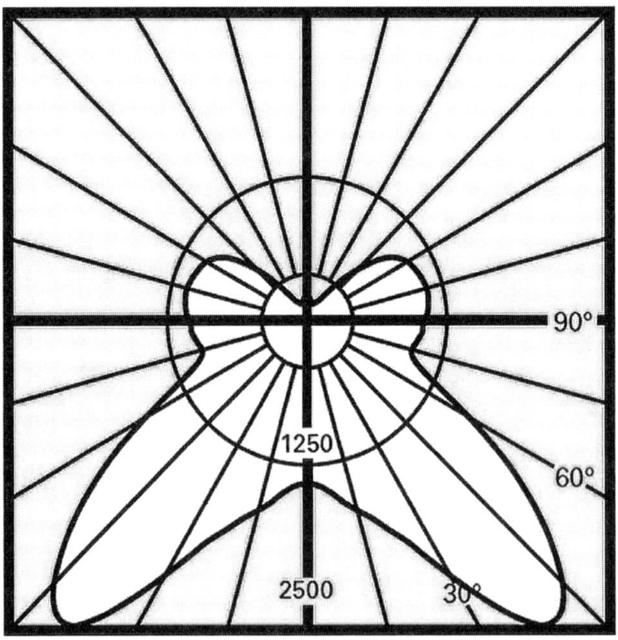


Candlepower

| Degree | CP |
|---|---|
| 0 | 1422 |
| 5 | 1454 |
| 15 | 1711 |
| 25 | 2125 |
| 35 | 3104 |
| 45 | 3233 |
| 55 | 2228 |
| 65 | 1251 |
| 75 | 972 |
| 85 | 1041 |
| 90 | 1058 |
| 95 | 1090 |
| 105 | 1112 |
| 115 | 1070 |
| 125 | 967 |
| 135 | 633 |
| 145 | 286 |

**Figure 3-8. Luminous Intensity Chart for an industrial high-bay fixture.**

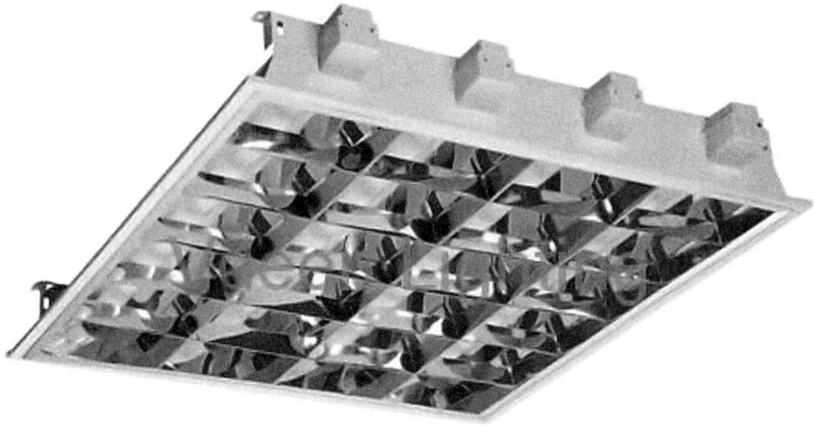

**Figure 3-9. A specular (highly reflective) louver used in a fluorescent parabolic troffer downlight.**

design with bare lamps shielded by a cellular louver. They are very efficient because they have no lens to absorb the direct light from the lamps and because the specular louvers are specifically designed to shield the lamps (to minimize the luminous intensity of the fixture at higher angles) while reflecting this shielded light down in a controlled fashion. These fixtures were so good at limiting luminous intensity at higher angles, and thereby controlling reflected glare on CRTs, that the ceilings and upper walls of the spaces in which they were used were often underlit in comparison to the horizontal surfaces of the work-plane and floor. The resulting *cave effect* brought about by the high level of contrast between the upper and lower surfaces made the environment feel dark and oppressively small. Thankfully, computer screen technology has progressed and the newer, low-glare flat screens have supplanted the old, curved glass video screens, making the use of parabolic lighting fixtures in open offices obsolete. Now we favor a combination of direct and indirect lighting designed to throw significant amounts of light up to the ceiling and out towards the walls. We refer to this direct/indirect approach as *volumetric lighting* because it reveals the volume of the space by lighting the ceiling and vertical surfaces. Volumetric lighting is a key concept for today's lighting designers, and one that we'll discuss in greater detail later on in this book. This style of lighting results in a much more open and airy feeling that is visually comfortable for the occupants even though it is predicated on the use of fixtures designed to project higher luminous intensities at higher angles. This is true because, as we've discussed, glare is relative, and as we brighten the surrounding surfaces we also raise our level of tolerance for brighter objects within the environment.

As is often the case it takes a while for trends to establish themselves, and we still see architects and engineers specifying parabolic troffers, even as the need for this type of fixture has greatly diminished. But we have also seen a new generation of recessed troffers emerge: the so-called direct/indirect basket troffer and the high-efficiency troffer. The direct/indirect basket troffer (Figure 3-10) is composed of a lamp, or multiple lamps, that are suspended below a curved, white reflector with a perforated "basket" that is in turn suspended under the lamps. Some direct light reaches the space below, through the basket, while indirect light is bounced into the space off of the white reflector. The high efficiency troffer (Figure 3-11) is of a similar design, utilizing a translucent diffuser in place of the perforated basket that allows for a greater degree of direct light while still making use of the reflector to create a larger, less glary, yet very efficient luminous source. Both of these fixtures project a greater luminous intensity above 55 degrees, but not so much as to create a direct glare condition, partly because they also tend to brighten the entire environment. In doing so they minimize the relative contrast between the fixtures and their surroundings by throwing a greater amount of diffused, ambient light onto the walls and other vertical surfaces. Also, because they are themselves larger luminous sources, they emit a lower level of luminance, as the amount of light from the lamps is spread out over a greater area. Figure 3-12 shows a representative luminous intensity chart for a fixture of this type. These fixtures create a more visually comfortable environment than the parabolic troffer, making them even more desirable as a workhorse for the millions of square feet of open office space utilized every day—especially where low ceilings are the norm. In spaces with higher ceilings a true direct/indirect fixture, usually a pendant designed to throw light upwards to the ceiling plane as well as down to the work-plane, will often be preferable.

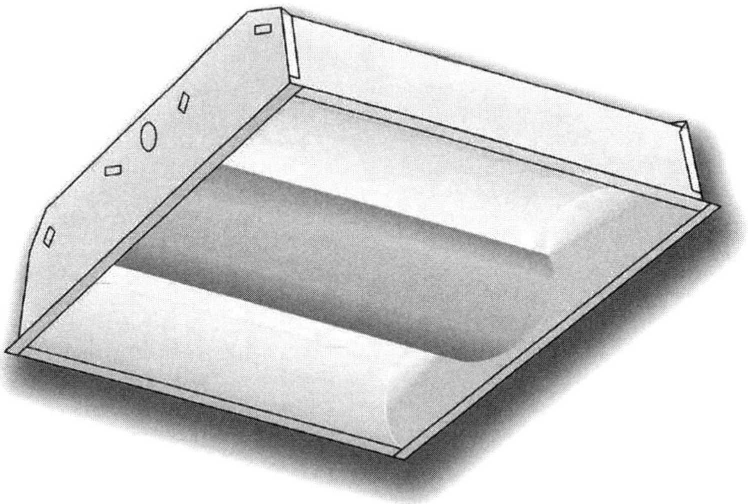

Figure 3-10. Direct/indirect basket troffer

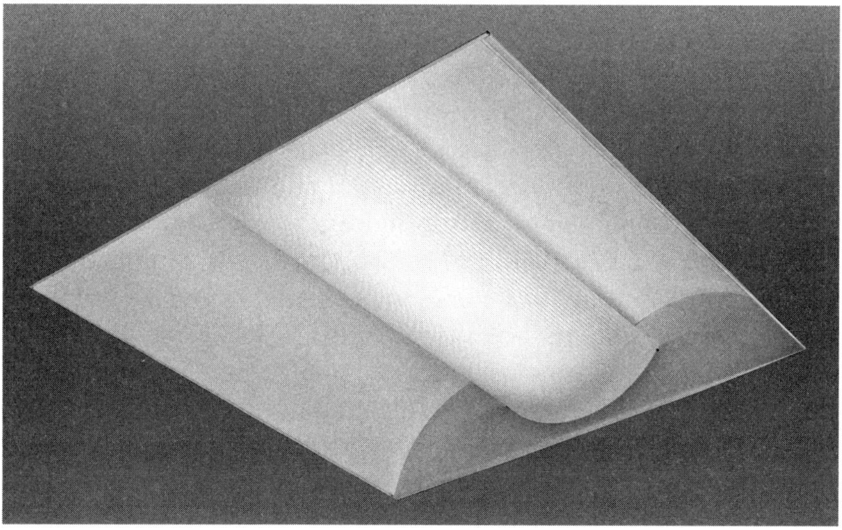

Figure 3-11. High-efficiency troffer

**CANDLEPOWER** DISTRIBUTION

| VERT. ANG. | HORIZONTAL ANGLE 0 | 45 | 90 | ZONAL LUMENS |
|---|---|---|---|---|
| 0 | 1218. | 1218. | 1218. | |
| 5 | 1211. | 1209. | 1205. | 115.4 |
| 15 | 1157. | 1160. | 1159. | 328.5 |
| 25 | 1052. | 1064. | 1074. | 492.5 |
| 35 | 893. | 935. | 960. | 584.8 |
| 45 | 703. | 772. | 827. | 595.7 |
| 55 | 510. | 605. | 689. | 539.3 |
| 65 | 321. | 449. | 555. | 441.2 |
| 75 | 156. | 307. | 419. | 313.3 |
| 85 | 26. | 80. | 103. | 79.8 |
| 90 | 0. | 0. | 0. | |

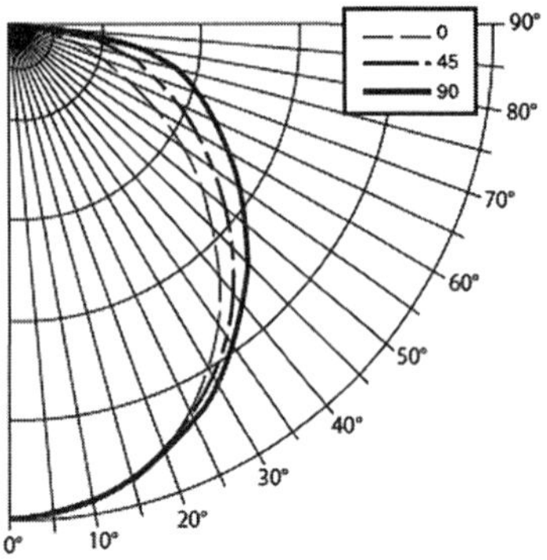


**Figure 3-12. Luminous intensity chart for a high-efficiency troffer**

## THE LUMINANCE FACTOR

Another way we can look at whether a lighting system is likely to cause glare is to consider the luminance of the light fixtures themselves, and as they relate to the luminance of the surrounding surfaces. Remember, luminance is a measure of the amount of light that is emitted or reflected back to the viewer per unit of area, in this case, candelas per square meter (cd/m$^2$). The IESNA has published some default recommendations for the maximum luminance of lighting fixtures in a number of generic commercial spaces, weighted by the importance of VDT-dependent tasks (critical, high, normal, and secondary) within each general category.[2] These range from a relatively low maximum luminance recommendation for lighting fixtures in VDT critical applications, like air traffic control centers, to a much higher maximum luminance for fixtures in industrial spaces where VDT-dependent tasks are of secondary importance. But the luminance of lighting fixtures is not easy to calculate, and it is not a metric that is published by many lighting manufacturers. In any case, these are general recommendations that need to be interpreted within the context of the actual environment in which the lighting is being installed. It is best to think about luminance in terms of contrast ratios between different surfaces, just as we looked at the luminous intensity of a fixture and how it relates to the surfaces around it. A pendant luminaire with a high-luminance diffuser will be more likely to cause glare when it is viewed against an unlit, dark ceiling than when it is viewed against a well-lit, light-colored ceiling. As we discussed in the last chapter, luminance is most closely related to our perception of the brightness of an object or surface. And since glare is relative, one way to minimize the glare from lighting fixtures is to create higher luminances on the surrounding surfaces, thereby reducing the luminance ratio, or contrast, between these surfaces and the luminaires in our field of view. One way we can achieve this is by specifying direct/indirect fixtures that illuminate the ceiling as well as the task plane, and by increasing the reflectance of the ceiling, as shown in the accompanying image (Figure 3-13) of a large, high-ceilinged waiting area. In low ceiling applications, specifying wall washers that illuminate the walls is another way to bring up the overall luminance of the various surfaces of the room, to minimize any glare effect from the overhead lighting.

**Figure 3-13. A waiting area lit with direct/indirect basket troffers and pendants. Image courtesy of Corelite division of Cooper Lighting.**

## MINIMIZING GLARE—IT'S NOT JUST ABOUT THE LIGHTING

In the end, it's impossible to adequately consider the implications of glare and contrast outside the context of the space or environment being lit. Are there windows? What's the ceiling color? Are the walls made of a specular material, like a glossy laminate, marble, or some other polished stone? Some spaces lend themselves to being lit in a visually comfortable fashion, and some, like the one in Figure 3-14 present greater difficulties. As with many of the factors we look for in a quality lighting design, managing contrast and controlling glare can be easier to do in some instances than others, and sometimes the biggest decisions affecting the lighting quality of a space are made before a lighting designer is even consulted. This is why lighting designers should always advocate for an integrative approach to design. Decisions ranging from a space's volume to its interior finishes should be considered along with an understanding of the occupants' visual requirements and the available means, including the lighting design options, to fulfill them. In this way we can best satisfy the aesthetic vision of the architects and designers and support the human needs of the occupants who will live and work in the buildings and interior spaces we are called upon to light.

**Figure 3-14. A poor choice of lighting fixtures in a difficult-to-light space creates a disorienting pattern of reflected glare combined with direct glare in a marble corridor. Photo Credit: Flickr user Oimax, [CC-BY-2.0 (www.creativecommons.org/licenses/by/2.0)], via Wikimedia Commons**

# Chapter 4

# Visual Comfort & Visual Interest

## WHAT IS VISUAL COMFORT & VISUAL INTEREST?

As we've already discussed, in a visually comfortable environment there is a sufficient quantity of light to reliably complete tasks, glare is minimized, and the luminance, or brightness levels, within our field of view are held within a certain range. Many factors will contribute to the visual comfort of an indoor environment including the number and position of windows, whether or not there are window treatments, the color and specularity of the surfaces, the size and shape of the room, the positions of the occupants within the space, and of course the lighting design itself. Visual interest is a phrase we use to describe lighting features that attract our attention, delight us, stimulate our visual sense, and provide for luminous variations. Visual interest may convey a mood, reinforce the shape or form of a space or building, or even tell a story. With lighting, the vocabulary at our disposal is mainly one of intensity and color. We create visual interest with light by varying the amount of light on different surfaces in a space, delineating new shapes or reinforcing existing shapes with highlights or shadows, and by manipulating the color of the environment. Interestingly, the qualities of light which usually enhance visual comfort—lower contrast ratios and subdued highlights—when taken to extremes will work against us when the aim is to create visual interest. Balancing the tension between these competing (and equally important) elements in lighting design is then a key part of the lighting designer's task. In general, we strive for even illumination on surfaces, or at least smooth transitions between areas of differing brightness, and we try to avoid variances in the amount of light in task areas where the occupants are engaged in similar activities. It can be tiring and disorienting to have your eyes constantly readjusting as you go about your day. On the other hand, if the brightness of all surfaces is too much alike, that environment can be lacking in visual interest. In addition, having a little more light on our work, or on a displayed object, than on the surrounding surfaces, can help focus our attention in a desirable way. So, while we usually design to keep the treatment of a given surface lit to the same brightness, we also look for opportunities to create visual interest with a complementary rhythm of lighting levels that may change from surface to surface, or from architectural detail to architectural detail. This technique can also serve to help define, outline, and accentuate the architectural form of the space we are lighting.

## THE OBSESSION WITH HORIZONTAL FOOTCANDLES

All too often, the single criterion used when developing a lighting layout is the quantity of light, or the illuminance, generated on the work-plane or at the floor. This measurement is referred to as horizontal footcandles: the amount of light falling on a horizontal surface with the light meter facing up, towards the ceiling. But does the light always have to be a direct source, coming from above? Because the sun is in the sky and most of our interior lighting is in the ceiling the vast majority of light is typically directed downwards. But do these scenarios produce the best lighting? The fact is, we see less well on a bright, cloudless day than we do when there is a layer of cloud cover. When the sky is clear and the sun is out, the direct light can be blinding, with a great deal of contrast between the lit areas and shadows. As soon as the clouds cover the sky, the light becomes diffuse, and at times almost directionless. It reflects off buildings, and is refracted through clouds, increasing the vertical footcandles (light falling on the sides of objects), and resulting in a nearly shadow-free light. Keeping this in mind, it makes sense that we would generally strive to illuminate the ceiling and vertical planes of our interior spaces. In doing so, we avoid creating a cave effect where there is adequate light on the floor and work-plane (horizontal footcandles) but the walls and ceiling are dark and the room lacks a bright and airy feeling. As mentioned in the previous chapter, we call this approach, in which we light the walls and ceiling to create an indirect return of light to the space, *volumetric lighting*. It is lighting that illuminates the bounding surfaces of a room and in doing so reveals the volume of the space. With this approach we increase the vertical footcandles to create a softer visual environment, with minimized shadows, in which it is easier to see the details of objects and the faces of our co-workers.

**Figure 4-1. Recessed linear cove lighting provides for a visually comfortable environment by illuminating the ceiling and floor, and creates visual interest by outlining and accentuating these curvilinear display walls. Photo & Lighting Design: Michael Stiller Design.**

## VISUAL COMFORT AND VISUAL INTEREST: HOW DO WE GET THERE?

The defining of metrics to help create a visually comfortable environment is an evolving art. In 1966, Sylvester K. Guth, a pioneer in the field of lighting, published a paper entitled *Computing Visual Comfort Ratings for a Specific Interior Lighting Installation* in the journal *Illuminating Engineering*, which described a process of mathematical calculations by which one could determine the Visual Comfort Probability (VCP) of a lighting design in a theoretical environment. In essence, the VCP value is a number that corresponds to the percentage of people that would find a particular lighting design, utilizing a specific fixture type, visually comfortable. The article demonstrated calculations that could be executed from a single point of view in a room with a proposed lighting layout. The procedure assigned a discomfort index for each luminaire, and then assigned an overall rating to the lighting system, taking into account many factors including room size and the surface reflectances within the observer's field of view. VCP calculations represent an early method used to predict the likelihood that a particular luminaire type, (especially a direct, recessed fluorescent fixture) would create uncomfortable glare for the occupants of a theoretical room with a regular layout of lighting and workstations.

As we have discussed, a variety of methods have since been developed to predict whether visual comfort (or discomfort) will be the likely result when using specific fixture types in specific applications. These include the establishment of maximum luminous intensities and luminance values, at certain angles, for lighting fixtures in different work-place scenarios. But to fully evaluate the degree of visual comfort and visual interest we will achieve, we need to consider the overall environment, and all of the surfaces within it, as a collection of luminances. It is, after all, the luminance of the surfaces around us that constitute what we actually see. Unfortunately, luminance calculations are inherently difficult to perform. Add to this the fact that luminance can be difficult to predict where specular surfaces are present, and the task becomes even harder. The behavior of light reflecting off of a specular surface, like a mirror, is significantly different than its behavior with a matte white wall, which while not specular is still very reflective. When dealing with specular surfaces, we need to keep in mind the *law of reflection* that states for specular reflections *the angle of incidence equals the angle of reflection.* The angle of incidence is the angle between a ray incident on a surface and a line perpendicular to the surface at the point of incidence. (See Figure 4-2 for an illustration of this law). This means specular, glossy, or mirror-like surfaces will create different luminance values from different viewpoints. (Think of looking at an oil painting in an art museum, and having to move around to find a viewing angle uncompromised by the reflected glare of the lighting fixtures or windows in the room.) This is an extreme example, but the color and finish of each surface, the angle from which it is lit, and the angle at which it is viewed will all work together in a complex way to affect that surface's luminance, and the viewer's perceived brightness of it. The variables can quickly multiply, requiring these calculations to be performed hundreds or thousands of times. Thankfully, we now have computer software capable of showing us representations of our lighting designs before they are installed, which can also calculate the illuminances and luminances that will result. With a properly "built" software model that includes the right input to define the reflectance and specularity of the various surfaces, and the proper luminaire data (usually supplied by the manufacturer in the form of a photometric report or an IES file that can be read by the various lighting software packages available today), it is relatively easy to calculate and predict these values for a given environment. And in doing so we can also create a quality rendering that accurately illustrates the resulting interaction of light. (See Figures 4-3 and 6-8 for examples of computer renderings that show the luminance of the surfaces and objects in a particular setting. Figure 4-3 does not account for the specularity of the various surfaces, while the rendering in Figure 6-8 does.)

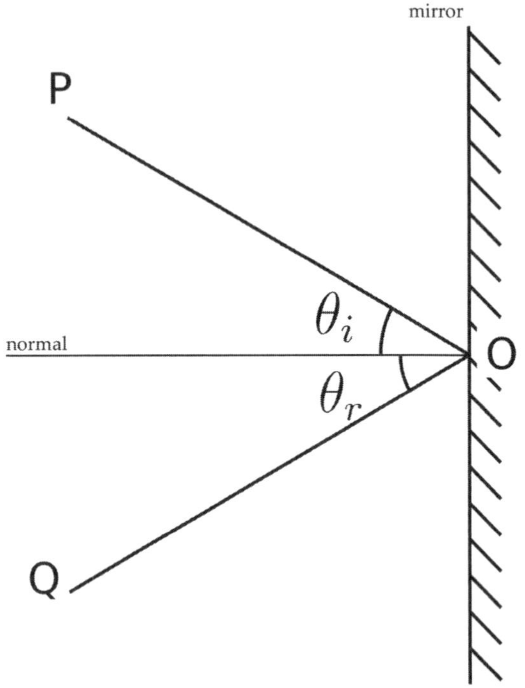


**Figure 4-2. A light ray PO strikes a vertical mirror at point O, and the reflected ray is OQ. By projecting an imaginary line through point O perpendicular to the mirror we can measure the angle of incidence, θi and the angle of reflection, θr The law of reflection states that θi = θr, or in other words, the angle of incidence equals the angle of reflection. Image Credit: Johan Arvelius, [http://creativecommons.org/licenses/by-sa/3.0/deed.en)], via Wikimedia Commons**

**Figure 4-3. Luminance calculations displayed as a grey-scale gradient. Image by AGI32 software.**

Given that lighting design is an art as well as a science, seasoned designers rely on previous experience as much as software models, or other calculation methods, when thinking about how to increase the visual comfort and visual interest of a space. For a higher degree of visual comfort, it is often enough to follow the guidelines we have already discussed: use indirect lighting to illuminate the ceiling and walls whenever possible and consider the placement of your fixtures in relation to the kinds of surfaces you are lighting. To achieve a sense of visual interest, we try to vary the intensity of the lighting in subtle ways that echo or follow the architectural lines of the space. And we always keep in mind the specularity of the various surfaces being lit. A cove lighting treatment may work well with a ceiling finished with a reflective, matte surface; whereas glossy paint can create enough specular reflection so as to reveal an image of the lamps or sources concealed in the cove in an unintended and undesirable way. Similarly, a task light over a specular work surface may create as much reflected glare as useful work light if it is positioned improperly, or if it is not equipped with a lens to redirect the light away from the viewer. (See Figures 4-4 and 4-5.) These are common pitfalls, and in most cases detailed computer renderings or calculations are not necessary in order to identify them. They are the result of everyday phenomena familiar to any keen observer of light. Paying attention to our memory of the way light interacts with various surfaces, and thinking through our design choices, will go a long way towards assuring the quality of our lighting designs. But when lighting large areas with predominantly dark or shiny surfaces, or other unusual environments, we may have to use software and other calculation methods to take a closer look.

## LUMINANCE RATIOS

While there is some controversy about how much contrast or how great a range of luminance should be allowed between

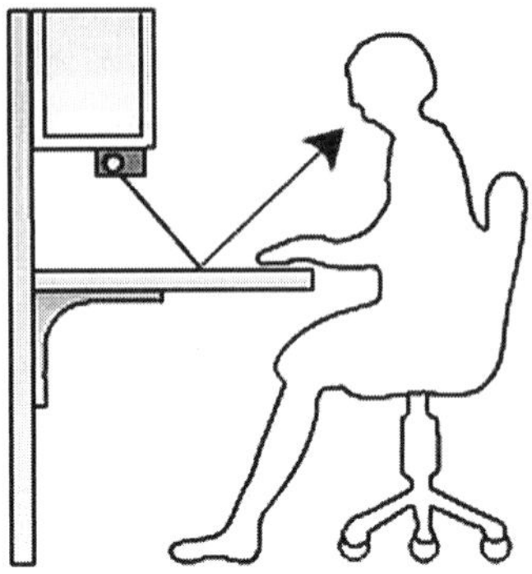

**Figure 4-4. Task lights that do not redirect the light away from a person's view can cause veiling reflections and/or reflected glare. From the IESNA Lighting Handbook, 9th Edition with permission from the Illuminating Engineering Society of North America.**

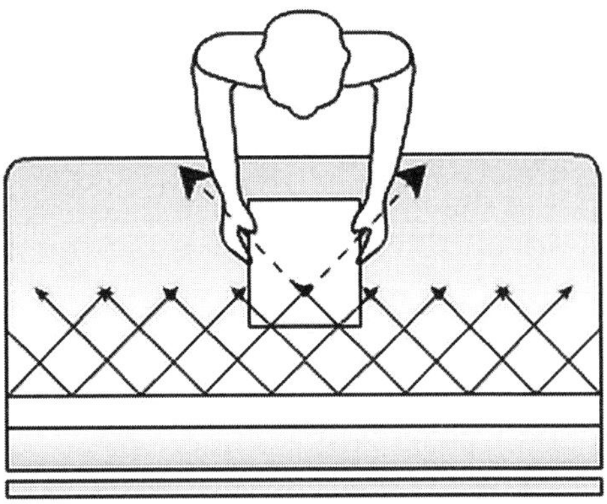

**Figure 4-5. With an appropriate lens, light from task lights is redirected to help eliminate veiling reflections and/or reflected glare. From the IESNA Lighting Handbook, 9th Edition with permission from the Illuminating Engineering Society of North America.**

interior surfaces, it is generally agreed that much greater contrast ratios may be tolerable, or even desirable, in some environments. Too much contrast can be perceived as creating glare, and too little as lacking in visual interest. An exposed lamp may be viewed as a glare condition in one environment, and an added bit of sparkle in another. Clearly, the desired degree of contrast in an environment can be very subjective, depending on the individual and the setting. A romantic restaurant, in addition to being less brightly lit overall, is likely to contain greater contrast ratios than an open office or an airport. The amount of contrast in lighting design can be part of a statement about the identity of a particular retailer or hospitality brand. We can even identify general tendencies in the contrast of lighting designs overall that follow a particular fashion or trend, changing from year to year. For example, in recent years the traditional, higher contrast lighting designs favored by high-end retailers, in which a somewhat dim ambient environment is punctuated by accent lighting to highlight products or architectural features, has been supplanted by cleaner, modern designs that favor larger, featureless surfaces with perfectly even illumination. But though we can manipulate an environment's luminance ratios for an emotional or visual effect, there are a few rules of thumb we should keep in mind when designing lighting for work spaces and other public environments where it is important that people of different ages, and with different visual abilities, can see well and function effectively.

Most studies of luminance ratios have concentrated on workplace environments, where the occupants are engaged in

**Figure 4-6. Luminance measurements in an office environment, expressed in cd/m2(illuminance measurements in italics). The luminance ratios are higher than what is recommended for a visually comfortable working environment: over 20:1 between the paper copy board and the window behind it. B. Piccoli, G. Soci, P. L. Zambellli, D. Pisaniello, Photometry in the Workplace: The Rationale for a New Method, Annals of Occupational Hygiene, 2004, Vol 48, Issue 1, by permission of Oxford University Press**

the same tasks for many hours at a time, and fatigue and eyestrain can lead to decreased worker productivity. The 10$^{th}$ edition of IESNA's The Lighting Handbook, considered by many to be the lighting designer's "bible," recommends the following as a general guide for default luminance ratios: [3]

Task to immediate background surfaces = 3:1
Task to dimmer distant background = 10:1
Task to brighter distant background = 1:10
Task to light source (windows or luminaires) = 1:40
Adjacent surfaces to light source
(windows, skylights, or luminaires) 1:20

It is generally accepted that higher luminance ratios are more likely to be tolerated, or even preferred, when they are visually justified, as in the highlighting of an object of particular visual interest, or where these variations in luminance levels echo architectural features. It is also generally agreed that in the immediate vicinity of a worker's task area, lower luminance ratios are desirable. Figure 4-6 shows an example of a very high luminance ratio, over 20:1, between a task area (the paper copyboard) and adjacent background (the windows) that would likely cause discomfort and diminish the productivity of any worker.

## LUMINANCE RATIOS AND DAYLIGHT

Another important consideration when examining likely luminance levels and contrast ratios in an interior environment is the quality and degree of daylight infiltration, especially where daylighting is part of the design program. Daylighting, which we will cover at greater length in a later chapter, is the practice of placing windows or other apertures in relation to reflective surfaces such that, during the day, natural daylight provides *effective* interior illumination. Daylight infiltration simply refers to the phenomenon of daylight entering a space through windows, skylights, and other apertures. So it follows that daylighting is a technique for maximizing *and* controlling daylight infiltration to create useful interior lighting while minimizing any potential for glare. Most people prefer an office with a window that provides a view and lets in natural light. Natural daylight is also believed by many to be beneficial in many ways, both psychological and physiological. And the use of daylight to illuminate interior environments can be a key component of lighting energy-efficiency, augmenting, or even replacing, the use of electric lighting at some times and in some situations. But daylight is also extremely bright. Even though higher contrast, or higher luminance, ratios are often tolerated in daylit spaces (thanks to our natural tendency to prefer being near a window) the difference between the amount of light in an interior space and the exterior environment can be great indeed. The average interior illuminance in most electrically lit spaces falls between 10 and 60 footcandles, and the exterior illuminance at noon on a cloudless day is about 10,000-12,000 footcandles. That represents between a 166 and 1,200-fold difference in the amount of light between the interior and exterior environments, which can translate into very high luminance ratios between surfaces lit by daylight, and those lit by electric lighting. Clearly a major challenge with daylighting is making use of all that free sunlight without creating extremely high luminance ratios between the windows and sunlit walls, and the rest of the interior surfaces. And so it follows: if the effects of daylight infiltration are not considered, or if poorly conceived daylighting treatments are employed, extremely high luminance ratios can unintentionally result in a serious problem with glare.

# Chapter 5

# Color & Light

## COLOR TEMPERATURE

Color temperature is a measurement used to define the different hues of white light sources. The scale, measured in degrees Kelvin, is based on the temperature of a theoretical object called a black body radiator, which when heated, will glow—much like the heating elements in a toaster. When heated to a lesser degree it will glow with an orange color. As it is heated more it will begin to glow with a whitish color, eventually turning bluish-white, like molten steel in a foundry. Though we refer to orange colors as warm and blue colors as cool, in the context of our color temperature scale the higher, or hotter, temperatures produce the cooler colors.

At this point it might be advantageous to step back and ask, what is white light? If you've ever played with a prism you know that what we refer to as white light is really an amalgam of light from the entire visible spectrum, literally comprised of a rainbow of colors that, when mixed together, create white. White light does not really refer to a specific color of light. Rather, it exists along a continuum called the *Planckian locus*. A locus is a set of points that satisfy certain criteria, and the Planckian locus is the set of points in our color space that follows the path of a black body radiator as it is heated. It is also a path through the space where all the visible colors of light come together and mix to create white. Depending on the exact composition of this "white" light, it might be a little warmer or cooler, a little more amber or perhaps a little more blue. Figure A-1, in Appendix A, shows an example of the Planckian locus in the context of a choromaticity chart representing our standard CIE color space.

In the natural world different atmospheric, geographic, and time-of-day conditions can create different "colors" or color temperatures of white light. For example, overcast skies will filter the light from the sun and create a cooler, or bluer daylight than what we might see with direct sunlight at dawn. In our built environment, the different electric light sources we use can also create different color temperatures. We can often choose from a wide range of "white light" color temperatures by specifying different lamps, or in the case of LED lighting different fixtures. As previously noted, we define this color temperature, also referred to as correlated color temperature (CCT), according to a numerical scale measured in degrees Kelvin. The color temperature of a particular white light is the same as the physical temperature of our black body radiator when it is heated to the point where it glows with that color. We denote color temperature with a number (the number of degrees) followed by the letter "K," for Kelvin. As we already discussed, this can seem counterintuitive, as with this scale the greater the number of degrees, i.e. the hotter the black body, the "cooler" the color of the light produced. So a warm light source, like an incandescent lamp, might have a color temperature of 2,700K, and a cool fluorescent lamp might have a color temperature of 4,000K. Despite the popular misconception, natural daylight is actually cool, or bluish, in comparison to most of our interior lighting, with a color temperature of approximately 5,500K-6,000K. Figure A-2, in Appendix A, shows a color temperature chart with various lighting sources compared to natural daylight at different times of the day.

With many electric light sources available in different color temperatures, the color of our interior light can be tailored to suit the requirements of the occupant, or the use of the space. The color temperature of white light has universal emotional resonance. Warmer light has a more intimate feel, and is usually desirable in residential applications, restaurants, and hospitality environments. Cooler color temperatures are often more appropriate for offices and many commercial settings. Another consideration may be the degree of daylight that is incorporated, via skylights and windows, into the interior space. If a significant amount of daylight contributes to the interior lighting during the day, the electric lighting should be cool as well if the desired result is for these sources to blend together. Conversely, an accent can be created, as a design statement, by using warmer electric light sources in conjunction with daylight to create areas of color contrast. Figure A-3, in Appendix A shows a small sampling of warm and cool color temperature lamps.

## CRI & COLOR QUALITY

CRI, or color rendering index, which should not be confused with correlated color temperature, is a metric used to gauge the ability of a specific light source to accurately render the color of an object to the viewer. Many of us have had the experience of buying something, such as an article of clothing, only to find it looks quite different once we get it home or when we look at it under daylight. Or perhaps the lighting in a particular environment makes our skin take on an unattractive hue. This is the kind thing that can happen with electric lighting of a low CRI. To establish the CRI of a particular lamp, or source, a series of eight color samples are viewed under the light generated by that source, and then under an ideal light source of the same color temperature. The results are then compared, and a rating is assigned to the source depending on how well, on average, it renders the eight color samples. The CRI of a lamp, or source, is rated on a scale of 1 to 100. Both daylight and incandescent (or halogen) are considered ideal sources, with a CRI score of 100.

CRI was developed around the middle of the 20th century when fluorescent and high intensity discharge lamps were enjoying wide adoption in commercial buildings, and scientists began to take an interest in assessing their ability to accurately render various colors. All good quality fluorescent and high intensity discharge lamps, and LED sources, should have a published CRI rating. But CRI is a controversial metric. While the CRI test procedures may predict the ability of a particular lamp or source to accurately render the eight samples *on average,* it is not a good predictor of a light source's ability to accurately render any one specific color. So if we have two sources with a CRI of 75, it is possible that one of them will do a poor job of rendering blue colors, and the other may do a similarly poor job rendering reds. Some researchers have found CRI to be inaccurate when evaluating LED sources, and the National Institute for Standards and Technology (NIST) is currently developing a new measure, called the color quality scale to take its place.[4] Even so, for the time being CRI is the only available metric to gauge the color rendering of a light source, and it is generally recommended that in any regularly occupied interior environment, lamps and light sources with a CRI of at least 80, and preferably 85 or higher, should be used.

# Part II

# How Much Light Do We Need & Where Do We Need It?

# Chapter 6

# Lighting + Space: Calculating the Results

## AVOIDING THE TENDENCY TO OVER-LIGHT

One of the key things we can do when designing an energy-efficient lighting system is to avoid the pitfall of over-lighting interior spaces. Engineers, interior designers, and architects who lack the proper tools to predict the illuminance that will result from their lighting layouts often tend to over-specify the wattage of lamps, as well as the number of lamps and fixtures they need in their reflected ceiling plans. This tendency to err on the side of caution is understandable. Not having enough light in an environment is a serious problem, and any owner would take their design team to task over it. On the other hand, having a little (or a lot) more light than required is not likely to raise as many eyebrows. So the safest approach for many is to round up when deciding how many fixtures to use in their layouts, or err on the side of safety when choosing the wattage of their lamps. This approach may be safe, but it's not very energy-efficient or sustainable, and it is not likely to help the lighting installation meet the restrictions of the local energy code.

## THE RESULT OF OUR LIGHTING DEPENDS ON MORE THAN OUR LIGHTING

It is important to avoid the pitfalls of playing it safe by over-lighting, but how do we know when our lighting fixture layouts will provide enough, but not excessive amounts of light? Let's begin to answer this question by talking a little bit about the way light works and interacts within an interior environment. We measure the amount of light in the various parts of a space in terms of illuminance. Remember, illuminance is the amount of light per unit area that falls onto a given surface. In the United States, the standard unit of measure for illuminance is the footcandle, which represents the number of evenly distributed lumens falling on one square foot of a given surface. (In Europe, and most of the rest of the world, the measure of illuminance is the lux, which is the number of evenly distributed lumens that falls on one square meter of a given surface.) So if we say the illuminance on a given surface is one footcandle, it's the same as saying that the amount of light falling on that surface is equivalent to one lumen per square foot of the surface area. Light can be concentrated, and the effect is cumulative. And when it is the illuminance will increase. It follows then that if the same number of lumens are projected from two different lighting fixtures, but one of them is designed to focus more of the lumens in a particular direction—down to the work-plane for example—then that fixture will provide a higher illuminance to that surface.

In addition to concentrating lumens we can also add them up from different sources, whether these sources are additional luminaires or are the result of inter-reflected light—i.e. light emitted from reflective surfaces that bounce the light from the luminaires back to our target surface. (See Figure 6-1 for an illustrated example of inter-reflected light.) In the case of an indirect lighting system, or even a direct lighting system in which some of the light strikes walls or nearby furniture, the color and reflectance of the walls and furniture will have an effect on the number of lumens—the illuminance—that will ultimately be directed onto the target surface. If the walls, ceiling, floor, or furniture are dark, they will absorb more of the light that initially strikes them. If they are light in color, they will reflect more light back, and in doing so add to the illuminance on the other surfaces in the room. In short, to calculate the illuminance we will achieve on any surface in a given environment we need

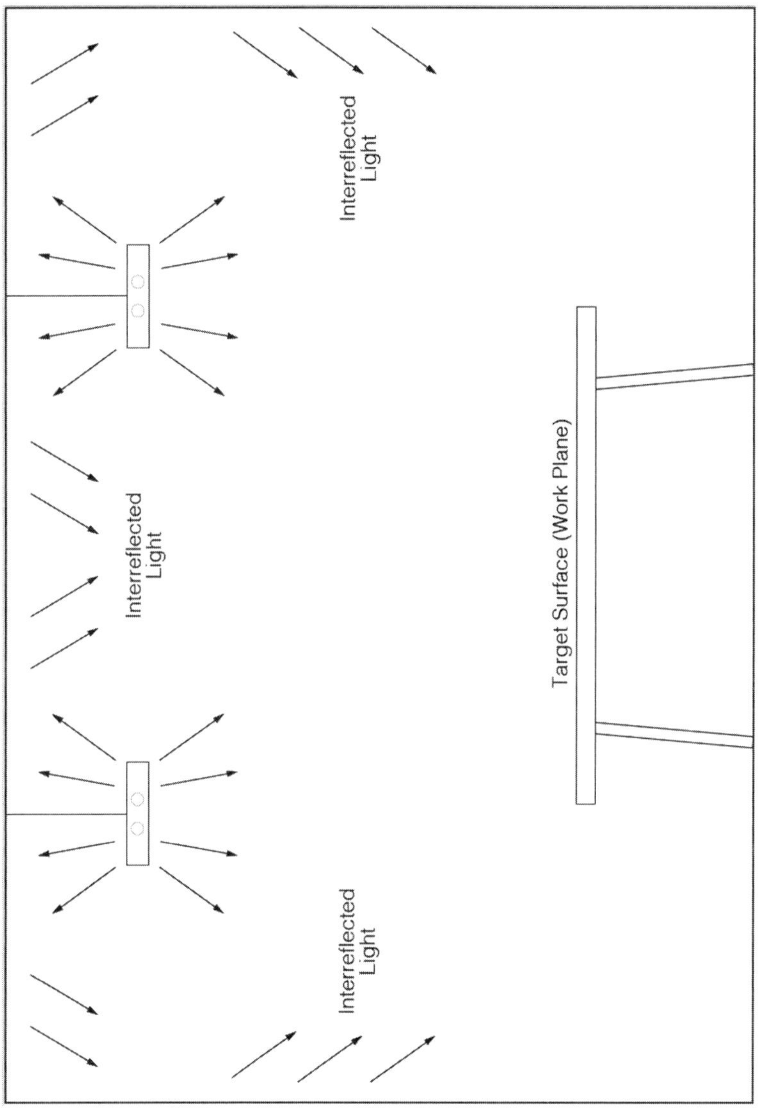


**Figure 6-1. Inter-reflected light, and the color and reflectance values of all surfaces must be considered when calculating the general illuminance in a room.**

to account for more than the direct output and layout of our luminaires. We also need to consider the luminous distribution of the luminaires (the amount of the light that is projected in each direction) as well as the reflectance of the walls, ceiling, floors, and even the furniture. A common example of this phenomenon is a system of wall-washers, which by their very nature will produce a distribution of light that falls mostly on the surface it is intended to light—in this case a wall—and not also strike the surrounding surfaces, like the ceiling or floor. If the wall-washers are directed onto a light colored wall, and the surrounding surfaces are also light in color, the inter-reflection of light may be great enough to generate an ambient illuminance at the floor level sufficient for someone to comfortably navigate their way through the space. Many passageways and corridors are lit today in this fashion.

## CHOOSING A CALCULATION METHOD

With all the variables to consider, calculating the amount of light, or the illuminance, on a particular surface in a given environment can seem like a daunting task. It is complicated, and there are many methods we can employ. We can perform calculations by hand, or use any one of the commercially available lighting calculation software products, some of which are listed in Appendix B of this book. Some luminaire manufacturers provide on-line calculators as part of their websites. And many manufacturer's representatives will provide a basic set of calculations specific to your fixture layout, with generic surface reflectance parameters, as part of the sales process, provided you are specifying enough of their products. But it's not often the case that one manufacturer will be used for all the different lighting treatments in a project. And not all projects are likely to fit a generic reflectance model. It will often be advantageous to be able to do lighting calculations yourself, or at least hire a lighting consultant to do it for you. Thankfully, there are a number of methods that can be employed, and they are not all so difficult as to be out of reach for most design professionals. Whichever you choose, having a good idea as to the amount of light that will be generated by your lighting system, and the resulting illuminance levels you can expect to achieve, is key information for any lighting specifier. Too little light to support the tasks at hand will be unacceptable, and too much is just wasteful. Depending on the requirements and scope of the project, and the degree of development required in a particular phase of the design, a simple inverse square law calculation to determine the amount of direct light that will be delivered to the work-plane or floor by a series of downlights may be sufficient. Or it might be necessary to run a full-blown radiosity solution on a 3D virtual model in which all the luminaires in a complex design—and all the reflectance values of the ceiling, walls, partitions and floor—are considered, for a more complete and accurate calculation. (Both of these methods will be described later in this section.)

## INVERSE SQUARE LAW

There are a number of formulas for calculating illuminance, and the inverse square law is one of the simplest. The inverse square law can be used to perform point-to-point calculations, especially useful in determining the illuminance delivered from a direct point source (like an adjustable halogen, metal halide, or LED accent light) onto a surface at a specific distance. This method is based on the concept, previously discussed, that the amount of light (or the lumens) emanating from a source spreads out as it travels, and the illuminance decreases proportionally as the distance from the source increases. According to the inverse square law, radiant energy (in this case, luminous energy, or light) reaching a point that is twice as far from its source will be spread over four times the area, with the result that for every doubling of the distance between the source and the surface being lit, that surface will receive only one-quarter the energy, or one quarter the amount of light. The inverse square law will not take into account the light that is inter-reflected within a space, from the walls or ceiling for example, and is only useful when determining the amount of light that will be delivered directly from the source. The formula to calculate illuminance using the inverse square can be found in Appendix B of this book. A graphic representation of the concept behind this mathematical law can be found in Figure 6-2.

## ZONAL CAVITY METHOD

Also known as the lumen method, the zonal cavity method is a hand calculation method for predicting the average indoor illuminance in a given space. The zonal cavity method is used with non-point source lighting (i.e. fluorescent and diffused

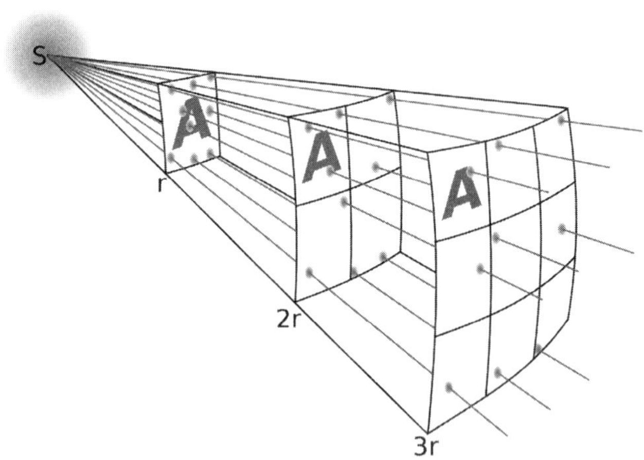


**Figure 6-2. Inverse square law. The lines in this diagram represent the luminous flux, or lumens, emanating from a light source. The total number of lumens depends on the output of the source and is constant. A greater density of lumens (lines per unit area) will occur at nearer distances, and the density of lumens (illuminance) is inversely proportional to the square of the distance from the source. Image Credit: Borb, [CC-BY-3.0 (www.creativecommons.org/licenses/by/3.0)], via Wikimedia Commons**

sources) and takes into account the various room surface reflectance values to give the average work-plane illuminance level for a room. This method provides a much better overall picture than an inverse square law calculation, as it takes into account the fact that lighter colored, and more reflective, walls and ceilings will result in higher work-plane illuminance, with more of the light redirected to the task area instead of being absorbed by otherwise darker surfaces. A conceptual outline of the zonal cavity method follows, and a detailed example of this method to calculate illuminance levels can also be found in Appendix B of this book.

The zonal cavity method is a series of steps and calculations whereby we first determine the effective reflectance of the different parts, or cavities, of the room by factoring in the room height as it relates to the work-plane and the height of our lighting fixtures. The room is broken down into three cavities:

1) The ceiling cavity (HCC), which is the space between the ceiling and the lighting fixtures. *If the lighting fixtures are surface mounted or recessed, the ceiling cavity ratio is 0 and the effective reflectance of the ceiling cavity is the same as the actual reflectance of the ceiling surface.*

2) The room cavity (HRC), which is the space between the work-plane and the lighting fixtures.

3) The floor cavity (HFC), which is the space between the work-plane and the floor. *If the work-plane is the floor, the floor cavity ratio is 0 and the effective reflectance of the floor cavity is the same as the actual reflectance of the floor surface.*

Once these cavities are defined it is possible to calculate numerical relationships between these spaces, called cavity ratios. The effective reflectance of the ceiling and floor cavities are then determined from a table that references their cavity ratios and the actual reflectance values of their surfaces (see Figure B-4 in Appendix B). Reflectance values are denoted as a percentage

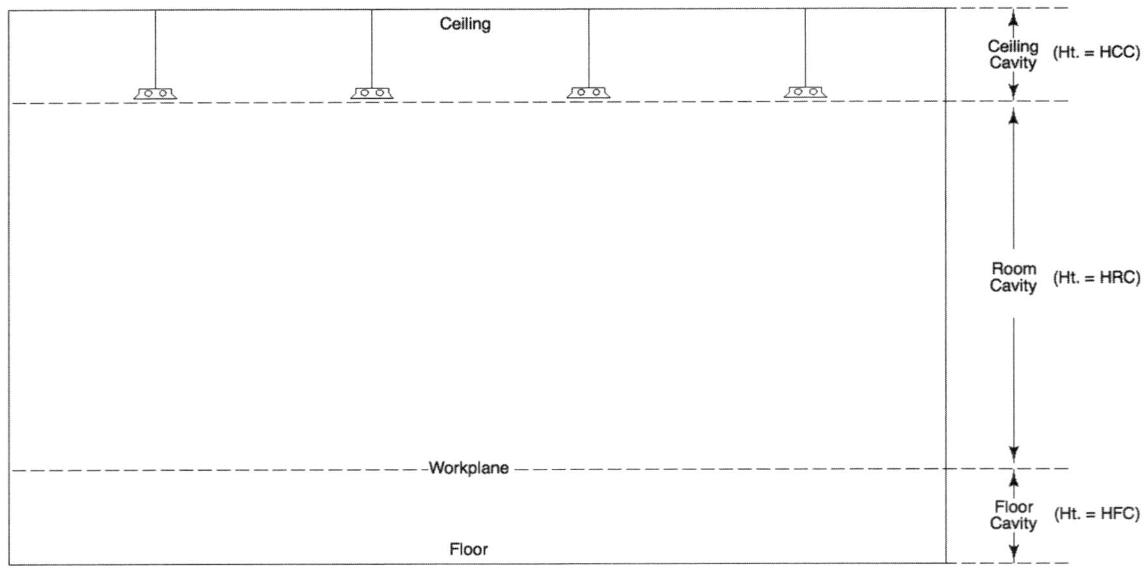


**Figure 6-3. Cavities of a room used to calculate average illuminance with the Zonal Cavity or Lumen method.**

from 0-100, with 100% reflectance being the value for an ideal material that will absorb no light, but reflect 100% back.

Next we find the coefficient of utilization value for our luminaire type from a table published by the lighting fixture manufacturer that references the wall reflectance and ceiling cavity effective reflectance values and accounts for the amount of light from the luminaire that will fall directly on the work-plane as well as that which will arrive there as a result of inter-reflections. A final adjustment to the coefficient of utilization may be taken from another chart in cases where the floor cavity has an effective reflectance other than the assumed 20%.

With the final coefficient of utilization determined, a calculation is then performed taking into account the layout and number of fixtures, the number of lamps, and the total lumens per lamp (including any applicable light loss factors from lamp depreciation, ballast inefficiency, or dirt buildup), to give the average illuminance in the room.

| RCC | 80 | | | | 70 | | | | 50 | | | 30 | | | 10 | | | 0 |
|---|---|---|---|---|---|---|---|---|---|---|---|---|---|---|---|---|---|---|
| RW | 70 | 50 | 30 | 10 | 70 | 50 | 30 | 10 | 50 | 30 | 10 | 50 | 30 | 10 | 50 | 30 | 10 | 0 |
| RCR | | | | | | | | | | | | | | | | | | |
| 0 | .86 | .86 | .86 | .86 | .73 | .73 | .73 | .73 | .50 | .50 | .50 | .29 | .29 | .29 | .09 | .09 | .09 | .00 |
| 1 | .78 | .74 | .71 | .68 | .66 | .63 | .61 | .58 | .43 | .42 | .40 | .25 | .24 | .24 | .08 | .08 | .08 | .00 |
| 2 | .71 | .64 | .59 | .55 | .60 | .55 | .51 | .48 | .38 | .35 | .33 | .22 | .21 | .19 | .07 | .07 | .06 | .00 |
| 3 | .64 | .57 | .51 | .46 | .55 | .49 | .44 | .40 | .33 | .30 | .28 | .19 | .18 | .17 | .06 | .06 | .05 | .00 |
| 4 | .59 | .50 | .43 | .39 | .50 | .43 | .38 | .33 | .29 | .26 | .24 | .17 | .15 | .14 | .05 | .05 | .05 | .00 |
| 5 | .54 | .43 | .38 | .33 | .46 | .38 | .33 | .28 | .26 | .23 | .20 | .15 | .13 | .12 | .05 | .04 | .04 | .00 |
| 6 | .49 | .39 | .33 | .28 | .42 | .34 | .28 | .24 | .23 | .20 | .17 | .14 | .12 | .10 | .04 | .04 | .03 | .00 |
| 7 | .45 | .35 | .29 | .24 | .38 | .30 | .25 | .21 | .21 | .17 | .15 | .12 | .10 | .09 | .04 | .03 | .03 | .00 |
| 8 | .42 | .32 | .25 | .21 | .35 | .27 | .22 | .18 | .19 | .15 | .13 | .11 | .09 | .08 | .04 | .03 | .03 | .00 |
| 9 | .39 | .28 | .22 | .18 | .33 | .25 | .19 | .16 | .17 | .14 | .11 | .10 | .08 | .07 | .03 | .03 | .02 | .00 |
| 10 | .36 | .26 | .20 | .16 | .31 | .22 | .17 | .14 | .16 | .12 | .10 | .09 | .07 | .06 | .03 | .02 | .02 | .00 |
| Floor Cavity Reflectance .20 | | | | | | | | | | | | | | | | | | |

**Figure 6-4. Coefficient of utilization chart for a specific luminaire that references the effective ceiling cavity reflectance (RCC), the wall reflectance (RW), and the room cavity ratio (RCR) with an assumed floor cavity reflectance of 20%.**

Although complicated, this is a relatively easy, if somewhat tiresome calculation to perform. However, it's not without drawbacks. The zonal cavity method will only take into account the contribution of one type of lighting fixture at a time, and only when regularly spaced in a regularly shaped room. In the end, the zonal cavity method is a good, though limited tool, especially useful in calculating the illuminance achieved from regular layouts of simple lighting treatments in spaces like open offices and warehouses. To make this process a bit easier, zonal cavity calculations can be performed with computer software. And some manufacturers include interactive, web-based zonal cavity calculators on their websites, like the one in Figure 6-5. These can be utilized to quickly determine the quantity, spacing, lamping, and wattage of a particular fixture type to use in a specific layout in order to achieve the general illuminance level required. In addition to a detailed example of a zonal cavity calculation, a list of some of the manufacturers who make available this kind of on-line calculator can also be found in Appendix B of this book.

## COMPUTER-BASED CALCULATIONS

Though the zonal cavity method can be an extremely useful tool to help the designer choose a layout for a single lighting fixture type in a regularly shaped space, there are many instances where the lighting requirements will be more complex. With many of today's lighting designs we rely on a combination of direct and indirect light from a variety of different sources, and from luminaires that are integrated into slots, coves, recesses, and other architectural features. The overall interior lighting achieved is the result of the many interactions between the various lighting treatments and these different architectural features, each of which may be finished so as to have a potentially different reflectance value. With the best of these designs, the result will be a system of luminaires and reflective surfaces, all work-

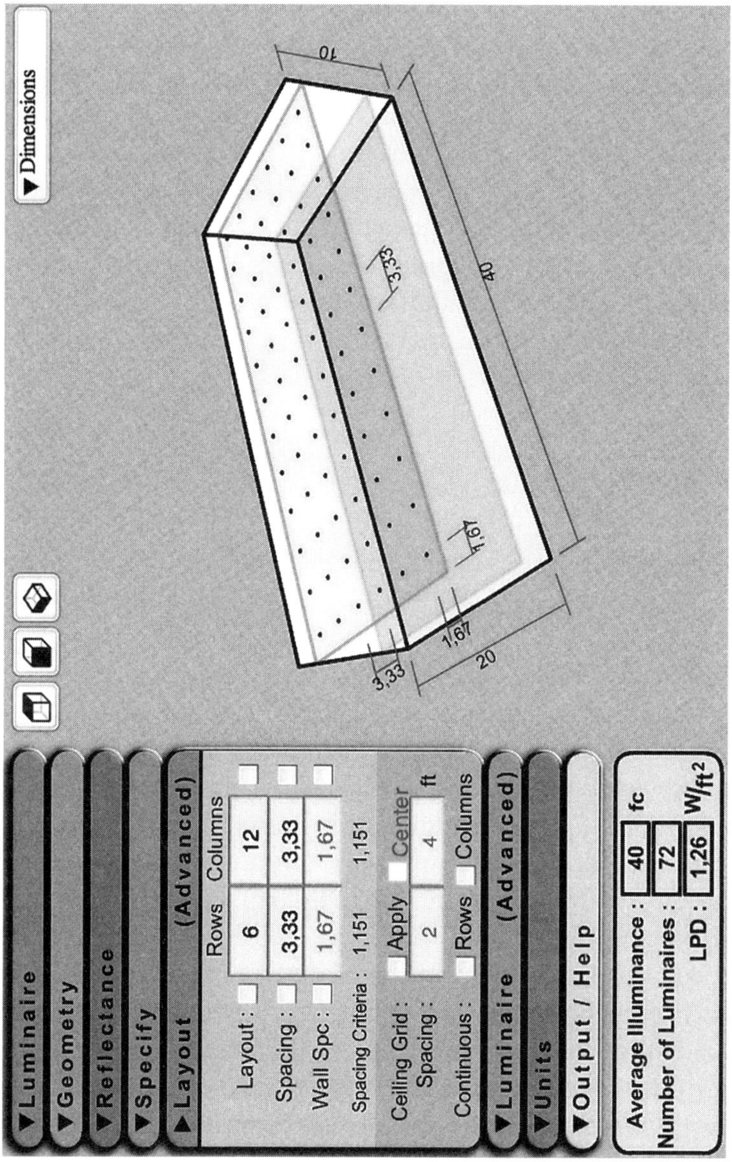


Figure 6-5. Online, interactive, zonal cavity illuminance calculator, Created by Flash_Indoor, copyright 2011 Lighting Analysts, Inc.

ing interdependently to produce a richly layered environment of luminances and illuminances. And with designs of this complexity, significantly more involved calculation methods are required to develop good predictions of the illuminance on the various surfaces, and the overall contrast, or luminance ratios, we can expect to achieve. Though designs with this degree of complexity may make it impractical, if not impossible, to perform useful calculations by hand, designers can make use of sophisticated computer software to analyze and depict the interaction of light that will result. And this type of software can generate photorealistic, three-dimensional renderings to help us effectively communicate the final look and feel of our lighting designs to our clients.

**IES Files**

At the heart of any computer-based lighting calculation is one or more small ASCII text files containing numerical information that defines the photometric performance of the luminaires being modeled by the rendering software. To standardize this process and insure as accurate a result as possible, these files should always be created as per the IESNA LM-63 Standard File Format for Electronic Transfer of Photometric Data and Related Information. Each lighting fixture type will typically have one of these IES files associated with it. They will usually have the computer file extension *.ies* (for the Illuminating Engineering Society) following the name of the file, which is commonly an alpha-numeric sequence containing elements of that fixture type's name and catalog number. Each modification to the lighting fixture that affects the output and distribution of light—the use of a semi-specular reflector in place of a matte white reflector, for example—will require the use of a different IES file. But some of the software programs that make use of these files also allow one to modify certain parameters without requiring the downloading and installation of an entirely new file. In this way a single IES file can be used to model the same fixture type with different wattage lamps, or perhaps different ballasts that may drive that particular lamp to a higher or lower output. Parameters can also be modified to model the light loss that will occur over time, as the lamps age or as dirt accumulates on the luminaire.

The way these IES files work from a practical point of view is relatively simple. First, the basic parameters of the space for which lighting calculations are to be performed are entered into the software program. With some of the more basic programs this can be done as a series of text entries whereby the size, shape, and reflectance values of the various surfaces within the space are defined. With the more advanced programs this is done by creating a three-dimensional computer model, in which a combination of keystrokes and mouse clicks literally draws a virtual version of the space on the user's computer screen. Again, reflectance values are defined for the walls, floor, and ceiling. Partitions and furniture can even be placed within the drawing and custom architectural treatments added, like dropped ceiling clouds, coves, and other details. Then the lighting layout is defined: the number of luminaires, their height, their spacing, and their focus (where appropriate). A specific IES file, representing the lumen distribution for an actual lighting fixture, is associated with each luminaire type. Within that file is all the data necessary to tell the calculation/rendering software how much light will be projected by each luminaire, and in which directions. These IES files are usually available from the lighting fixture manufacturers, sometimes as a simple download directly from their websites. And they can be interchanged to see the way the lighting will change if a different lighting fixture is used. In this way, a luminaire layout can be defined and the user may compare the performance of different fixtures, merely by substituting the data for another fixture type, or for the same type of fixture by different manufacturers.

**Radiosity & Ray Tracing**

A common calculation method used by many of the currently available lighting design software programs to create renderings is called *radiosity*. With this method, light is depicted in a very natural way, with soft shadows and gradient highlights. An example of a rendering generated with radiosity is shown in Figure 6-6. In a radiosity calculation the light from a source, be it a lighting fixture or the sun streaming into an open window, reflects from surface to surface until it is fully absorbed, illuminating the room along the way, just as it would in the real world. Radiosity calculations can be used to show the illuminance we can expect to achieve in a specific room at a specific place and time, taking into account the contribution of the electric lighting specified and the daylight that will be present, to give us a complete picture of the interior lit environment. Many of these software programs, which are for the most part easily run with the resources of today's personal computers, can also produce daylight studies to show the actual path of direct sunlight into a specific building or interior space. This kind of study can be of great assistance to architects and designers in developing appropriate window overhangs and light shelves, to bring more effective daylight into an environment while controlling glare. This can be an integral part of the practice of daylighting, which we will examine later.

The same radiosity method that creates these kinds of photorealistic images can also be used to generate a set of calculation points, as in Figure 6-7, that shows the illuminance anywhere in the room, again taking into account the daylight and electric lighting contributions and the interaction between this light and every surface it bounces off of. This method is by far the most widely used by lighting designers today to predict the illuminance levels they will achieve, and it can be an indispensible tool to aid in the design and specification of a lighting system.

Renderings based on a radiosity calculation can seem quite realistic, but they do not show us one important phenomenon that exists in the natural world: an accurate depiction of the interplay between light and specular, reflective surfaces. Recall that specular is a word that describes an object or surface that has the properties of a mirror. We may also use terms like glossy or shiny to describe specularity. We observe specularity in a vari-

**Figure 6-6. A 3D computer rendering generated by a radiosity calculation shows the interaction of light within an interior environment in a very naturalistic way. Image by AGI32 software.**

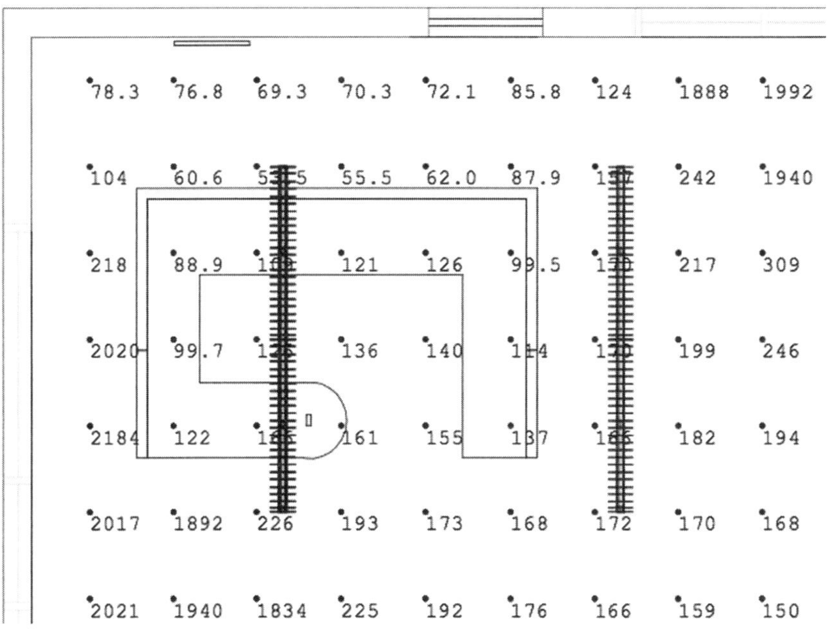


**Figure 6-7. Calculation points in a different part of the open office (depicted in Figure 6-6) showing illuminance values on a grid at the work-plane, usually 2.5 feet off of the floor. Image by AGI32 software.**

ety of materials, and in different degrees. For example, a glossy paint is not as specular as a mirror, but it will create a more defined reflection of a lamp—as a point of light, for example—than a matte surface. This phenomenon is lost in a radiosity calculation because, though our environment is full of materials and surfaces possessing some degree of specularity, a pure radiosity solution will treat all surfaces as diffuse (matte) reflectors, and will not consider any specularity they might possess. There is a practical reason for this. Many programs that create renderings based on a radiosity solution will render the entire environment in one process, such that multiple views, looking in multiple directions, can be generated at once. This process may take anywhere from a couple of minutes to many hours, depending on the complexity of the environment and the lighting being rendered, and the power of the computer used. As previously mentioned, with diffuse surfaces the luminance does not change from viewpoint to viewpoint in the same way as with specular surfaces. As anyone who has changed position to avoid glare knows, the reflection of light onto a specular surface will appear quite differently depending on the position of the viewer. So it follows that to render the light in an environment and include all the possible views, in all the possible directions, while considering all of the possible interactions between that light and any specular surfaces would take a far longer time to calculate.

There is another method of computerized lighting calculations which creates very realistic images that accurately show how specular objects will appear, even to the extent of simulating mirrors, or glassware, and showing us the reflections we will see in these objects. This method is called *ray tracing,* and it is especially useful for creating photo-realistic renderings to demonstrate a particular issue involving the placement of a lighting fixture in relation to a specular surface, to foresee any possible glare conditions or lamp reflections that might result, or simply to create as attractive an image as possible for presentation to a client. The process of completing a ray tracing solution can take quite a bit more time than a radiosity solution, and will be necessarily confined to a single view, or a series of defined views. Figure 6-8 shows a computer-generated image of glassware created with a ray tracing solution. These images can be breathtaking in their realism, and potentially valuable tools for marketing the lighting designer's craft, but it will not usually be necessary to use this method for everyday lighting design work. It should be employed, however, in cases where many, very specular surfaces are present.

**Figure 6-8. A computer generated 3D rendering created with the ray tracing method, converted from the original full color for the purpose of this book. An accurate calculation of the interaction between light and specular (shiny) surfaces results in a very realistic still-life of glassware in a sun-drenched interior scene. Image Credit: Gilles Tran**

# Chapter 7

# Target Illuminance Levels

## HOW MUCH IS ENOUGH?

A greener design uses less energy, and we can use less energy by providing the amount of light we really need and not more. So if we have a reliable way to predict the amount of light our designs will yield, the next question is, how much light do we actually need? This is often the first question asked when building design professionals consider their lighting designs. And while it's by no means the only important question, it is arguably one of the most important. But the answer is not easy to agree upon, nor has it remained constant for any sustained period. In fact, the more we examine this question, the more complicated the answer becomes. At least one study has shown that when given control of their own lighting, many individuals display a broad range of preferences for different illuminance levels, often choosing a level of light less than the current recommendations. (This finding also suggests that allowing individual workers to choose their own level of light, through the implementation of localized, manual dimming controls, would have the additional positive effect of reducing workplace energy use.) [5] So, the question of how much light is enough is a subjective one. And in addition to individual preferences there are other factors that have, over time, influenced our notion of what the right amount of light is, with the result that recommended lighting levels, or illuminance targets, have seen significant changes over the last century. These recommendations have changed as our technology has changed, as various economic and political factors have changed, and as our work environments have changed. There also seem to be cultural differences that influence the prevailing illuminance recommendations. It is interesting to note that illuminance targets have not only changed over time, but are in many cases significantly different in different countries during the same time period. All of this would seem to indicate that the central question with which we often begin the lighting design process is not one that can be easily answered by purely logical means.

## A MOVING TARGET

In the late 19th and early 20th century, the advent and commercialization of electric incandescent lighting, which cast a stronger and more reliable illumination than gaslight, made it much easier for workers to remain productive after daylight hours. In the ensuing decades, the desire for greater worker safety and higher productivity, combined with technological advances in lighting, resulted in the establishment of ever-increasing lighting levels in many workplaces. In the 1930's, with the introduction of the fluorescent and high-intensity discharge lamp technologies (sources that produced a greater amount of light per unit of energy), the cost of providing higher quantities of light for industrial and office environments fell dramatically, and recommended illuminance targets steadily increased. This trend continued until the early 1970's. But with the energy crisis of that decade and our increased awareness of the West's dependence on foreign suppliers of oil, along with subsequent changes in the way we work, the technology we employ, and concern for the environmental effects of burning fossil fuels to generate electricity, we have seen a steady decrease in recommended illuminance levels. In 1972 the IESNA recommended illuminance level for offices where VDT tasks were performed was 150 footcandles, and by 1993 these recommendations had fallen to 30 footcandles.[6]

**Recommended Maintained Horizontal Illuminances (lux) For Different Activities In Offices**

| Country & Year | General Lighting | VDT (computer) tasks | Reading | Drafting |
|---|---|---|---|---|
| Australia, 1990 | 160 | 160 | 320 | 600 |
| Austria, 1984 | 500 | 500 | --- | 750 |
| Belgium, 1992 | 300-750 | 500 | 500-1,000 | 1,000 |
| Brazil, 1990 | 750-1,000 | --- | 200-500 | 3,000 |
| Canada, 1993 | 200-300-500 | 300 | 200-300-500 | 1,000-1,500-2,000 |
| China, 1993 | 100-150-200 | --- | 75-100-150 | 200-300-500 |
| Czech Republic | 200-500 | 300-500 | 500 | 750 |
| Denmark | 50-100 | 200-500 | 500 | 1,000 |
| Finland, 1986 | 150-300 | 150-300 | 500-1,000 | 1,000-2,000 |
| France, 1993 | 425 | 250-425 | 425 | 850 |
| Germany, 1990 | 500 | 500 | --- | 750 |
| Japan, 1989 | 300-750 | 300-750 | --- | 750-1,000 |
| Mexico, proposed | 200 | --- | 900 | 1,100 |
| Holland, 1991 | 100-200 | 500 | 400 | 1,600 |
| Russia, 1995 | 300 | 200 | 300 | 500 |
| Sweden, 1993 | 100 | 300-500 | 500 | 1,500 |
| Switzerland, 1997 | 500 | 300-500 | 500 | 1,000 |
| UK, 1994 | 500 | 300-500 | 300 | 750 |
| USA, 1993 | 200-300-500 | 300 | 200-300-500 | 1,000-1,500-2,000 |

**Figure 7-1.** Recommended Illuminance Targets, as measured in lux, varied widely in different countries from 1984 to 1997. [6, 7]

Today the IESNA recommends target horizontal illuminance levels in office facilities that are anywhere between approximately 5 and 100 footcandles (expressed in the 10th edition of the Lighting Handbook as 50 and 1000 lux) depending on the primary use of the space and the visual requirements for the task being performed, and assuming the visual age of the workers is between 25-65 years. The lower end of this range is reserved for transitional spaces, and the upper end is the recommendation for security inspection areas. [8] (With this edition the IESNA has started publishing illuminance recommendations in lux, which is the metric standard. One footcandle is equal to 10.76 lux, and so it is easy to estimate the conversion from lux to footcandles by dividing the number of lux by ten).

Though these recommendations have varied from country to country, and have changed over time, a few things remain certain. We know that different people prefer different amounts of light, that different amounts of light are required for different tasks in different environments, and that people need more light as they get older. It may be impossible, or at least inadvisable, to design quality lighting based solely on standardized metrics. But it is nonetheless crucial to have a basic understanding of the illuminance targets that are generally considered appropriate for the different tasks, applications, and interior environments that we may be called upon to illuminate. To this end, the IESNA has developed a set of recommended illuminance targets that takes some of the more quantifiable variables into account and projects them into in a variety of default scenarios. These targets are necessarily based on the best and most current research available today, and will be predicated on a set of given conditions that may or may not actually exist in each individual project. The lighting designer's experience and judgment, developed over time and with sustained practice, will be required to factor in the differences between these scenarios and the real-world conditions that may be present in each project.

## RECOMMENDATIONS FOR TODAY

The IESNA Lighting Handbook, 10th Edition, establishes a wide-ranging set of criteria and recommendations for illuminance targets that can be used as a reference for specific types of applications, as well as a description of the general *illuminance determination system* applied to each. The illuminance determination system considers three basic factors that, when taken together, inform the illuminance recommendations for a particular task, age group, and setting. The illuminance target recommendations range from 0.5 lux for very low activity situations where a young observer's eyesight is adapted to a dark setting, like a night exterior, to 20,000 lux for health-care procedures where the observer is over the age of 65 years.

The three factors considered include:

- Task Characteristics
  - Size of and contrast between the objects and visible forms that are to be viewed in the performance of the task
  - Speed at which the task will generally be performed
  - Requirement for speed and accuracy
  - Enhanced ease of performance at low illuminance levels

- Task Importance
  - Is the task critical or secondary in relation to the other tasks to be performed in the same area or space?
  - Is the outcome of the task dire, or a matter of health and well-being? (medical procedures, emergency response, etc.)

- Observer Characteristics
  - Visual age (same as chronological age for normal-sighted individuals. When visual impairments affect an individual's ability to see, their visual age is greater than their chronological age.)

Task characteristics and task importance are factored together to create 25 categories (A-Y) of task characteristics such as *dark adapted situations, indoor commerce situations, indoor industrial situations,* and *health care procedural situations.* These categories are subdivided into five *visual performance descriptions* that include:

- Orientation, relatively large-scale, physical (less cognitive) tasks
- Common social activity and large and/or high-contrast tasks
- Common, relatively small-scale, more cognitive or fast-performance visual tasks
- Small-scale, cognitive visual tasks
- Unusual, extremely minute and/or life-sustaining cognitive tasks

Observer characteristics are then factored into the equation. For each of the 25 categories of task characteristics, three minimum maintained illuminance targets are indicated: one for situations where at least 50% of the observers are of a visual age of 25 years or less, one where at least 50% of the observers are of a visual age of 25-65 years, and one where at least 50% of the observers are of a visual age of over 65 years. [9]

The IESNA Lighting Handbook applies this system to a wide variety of applications, and in each case breaks the corresponding facility or project type down into various composite spaces and/or tasks. Three different illuminance recommendations are provided, taking into consideration the three categories of visual age of observers as outlined above. The recommendations go beyond simply looking at the horizontal illuminance on the traditional, desktop work-plane to also include:

- Vertical illuminance targets for each task or area (i.e. the illuminance that will fall on vertical surfaces, like walls, partitions, easels, and white boards)
- Uniformity targets (the ratio between the maximum and minimum variances in each area or target surface that should be allowed) for both horizontal and vertical illuminance targets, when appropriate

- Applications and tasks that are candidates for daylighting as a potential strategy to achieve the illuminance targets
- Applications and tasks involving specular surfaces (print with glossy ink or certain types of computer monitors, for example) where veiling reflections are a potential concern
- The area the recommended target illuminance applies to (the specific area where a task will be performed, or an entire room or designated space) [10]

Complete information on this procedure and the 17 different application types covered is available in the *IES Lighting Handbook,* which is an invaluable resource for any lighting designer. Application chapters include lighting for art, courts and correctional facilities, healthcare, libraries, offices, retail, residences, hospitality and entertainment, sports and recreation, and others. It is important to note that the general recommendations derived from a method such as the one advanced by the IESNA are a valuable set of metrics, but only as a starting point. Whether we are helping to forge an identity for a retail store or allowing workers to choose the lighting levels that suits them best, when it comes to lighting the best science will be that which is tempered by the art of our designs. And as we have seen, the science of quality lighting will continue to evolve, as it depends on a great many human factors that can, and will, continue to change.

# Chapter 8

# Task Lighting

## GETTING THE LIGHT WHERE IT'S NEEDED

In the previous two chapters we discussed the various methods to determine the illuminance, or the amount of light, we can expect to get from of our specified lighting designs, as well as a method to determine the target illuminance levels we should aim for. These are critical parts of the lighting design process. When specifying lighting for high performance buildings our aim must be to provide the appropriate amount of light for a particular task, or environment, while avoiding waste. We can best do this by not specifying lighting systems that generate more light, and therefore use more energy, than necessary. And we can take this concept one step further if we clearly define not just the amount of light we need for a given task, but where that task is likely to take place. We can then divide our space into task areas and circulation areas, specify an overall lighting system to allow the occupants to safely and comfortably move about, and then specify additional lighting fixtures to be installed in the task areas to provide higher levels of illumination where it is needed. In most offices the task area, or work-plane, is the workers' desktop. In an industrial facility it may be their workbench. In both of these examples there can be large areas that are mainly used for the circulation of the workers. This non-task area generally requires less light. The illuminance target examples from the last chapter state that the circulation areas in our offices might require as little as 5 footcandles (about 50 lux) of horizontal illuminance, whereas most task areas require between 30 and 50 footcandles. So, why light an open office as if tasks are to be performed in the entire area when those tasks actually occur in only parts of the overall space? The answer is: we shouldn't. Instead, we can employ a design technique, aptly named *task-ambient,* whereby we aim for lower ambient illuminance targets throughout the space and then supplement the task areas with lighting that raises the illuminance to the appropriate level when it's required.

A common task-ambient strategy is to employ indirect lighting, illuminating the walls and ceiling to create a sense of brightness and provide enough ambient illuminance for general circulation, and then specify a separate system of task lighting to increase the illuminance at the desks. A separate system of task lighting saves energy both by eliminating the requirement to light the entire space to a high level, and by creating the opportunity integrate local controls which allow the task lights to be dimmed by the users, and turned off when the work station is vacated. These controls can be manually operated or automatically triggered with occupancy sensors. Task lighting fixtures

**Figure 8-1. A high-tech, articulated arm design for an adjustable LED desk lamp allows the light to be directed to any part of the desktop. Image courtesy of Koncept Technologies.**

come in many forms, and can incorporate many different kinds of sources. Two of the most common types are surface mounted under-cabinet fixtures and adjustable desktop lamps.

## LOCALIZED GENERAL LIGHTING

In addition to providing dedicated adjustable task lights on workers desks, or integrating lighting fixtures within the furniture systems, there are a number of ways to create a task-ambient system by simply finessing the specification and layout of the permanently installed overhead lighting. As these are modifications to permanent installations of general lighting fixtures in the ceiling, and do not employ additional desk lamps, table lamps, or fixtures integrated into the furniture, they are inher-

**Figure 8-2. Fluorescent strip task lighting mounted to the underside of a cabinet is an easy and cost-effective method to provide greater illuminance on the desktop.**

ently less flexible and rely on the designers understanding of the layout of the space in order to be successful. But the advantage to these methods is that they are very cost effective, since they do not require the purchase, or wiring, of additional fixture types. The first of these techniques is to make judicious use of either direct or direct/indirect pendant lighting fixtures. We'll call it localized general lighting, and it's an approach that can work in both open offices and industrial spaces.

A direct/indirect pendant fixture is one that makes double use of the installed lamps by projecting some percentage of the lamp's output up toward the ceiling, and some percentage down to the floor and work-plane. When used with an appropriately reflective ceiling treatment, usually matte white or off-white, the

**Figure 8-3. A direct/indirect pendant fixture. These often come with specifications that indicate what percentage of the lamp is projected as indirect uplight, and what percentage is direct downlight. Some come with different diffuser and reflector options that enable the user to precisely select how much of the light is indirect and how much is direct.**

indirect light bounced back into the space from the ceiling will often be sufficient to create enough ambient illuminance for the purposes of general circulation. In addition, as it illuminates the ceiling it will have the effect of visually opening the space up. In lighting the perimeter surfaces we reveal the volume of the space, create a sense of overall brightness in the room, and greatly enhance the visual comfort of the environment. Along with the indirect uplight component of these direct/indirect fixtures, there is a direct, or downlight component that adds a higher level of illuminance to the work-plane below. By placing these fixtures right over our workstations, and at the appropriate height, as in Figure 8-4, we can create a task-ambient system with one fixture type in a very cost-effective way.

A similar approach can be used with all-direct lighting fixtures in which the ambient light is created with surface mount or recessed ceiling fixtures and the task light comes from direct pendants suspended at a lower level above the work-plane. We can vary the wattage of the lamps in each of these types, and even the number of lamps, to get the appropriate low level of

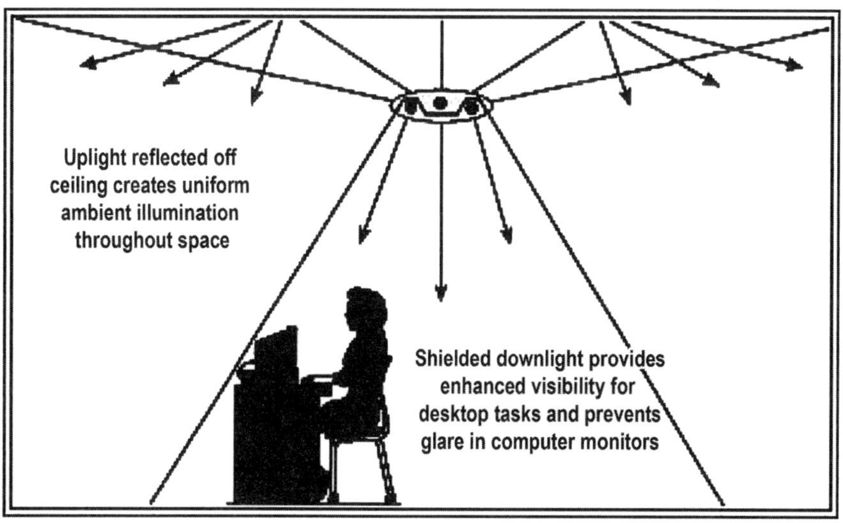


**Figure 8-4. A localized/general lighting scheme, utilizing direct/indirect pendant fixtures.**

ambient light for circulation, and a higher level for task lighting. This technique, shown in Figure 8-5, may be better suited to industrial settings, especially those with high ceilings. But since a direct/indirect system will almost always result in a more visually comfortable environment, we should specify one whenever feasible.

## FURNITURE-MOUNTED TASK-AMBIENT SYSTEMS

Another task-ambient system that combines both the cost-effectiveness of a localized general lighting scheme and the flexibility of portable task lighting is one that employs furniture-mounted direct/indirect fixtures. Like the localized general lighting scheme, this system also provides a complete task-ambient solution with a single fixture type, keeping costs for both equipment and installation as low as possible. And since the lighting fixtures are mounted directly to the furniture, they are easy to relocate in an open office as the workstations are reconfigured to satisfy the changing needs of the occupants. Figure 8-6 shows an open office utilizing these fixtures, which are mounted to the desks at a height that prevents direct glare from both the uplight and downlight component. The lighting fixtures contain lamps that project light up to the ceiling to create a visually comfortable, indirect, ambient light, as well as lamps that direct light down to the work-plane, to raise the illuminance to the required levels at the task area. Controls are easily integrated to provide manual switching, occupancy sensing, and dimming of the task lighting component for even greater energy cost savings. This system does rely on a highly reflective ceiling plane, and the effectiveness will depend to some degree on the ceiling height, with very high ceilings creating a greater challenge.

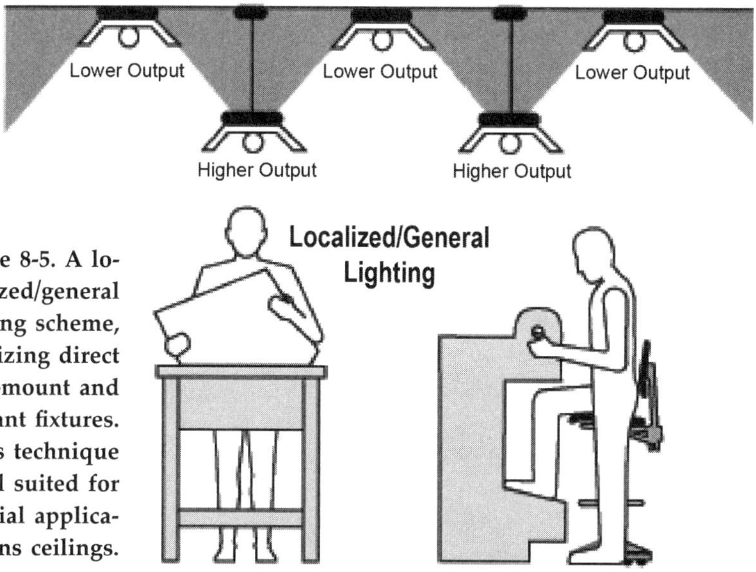


Figure 8-5. A localized/general lighting scheme, utilizing direct surface-mount and pendant fixtures. This technique is well suited for industrial applications ceilings.

Figure 8-6. Partition mounted direct/indirect fixtures create a complete task-ambient scheme with one fixture type. Photo credit: The Lighting Quotient, Bic Corporation, Shelton, CT., Architect: Fletcher Thompson

# Part III

# Sustainability & Electric Lighting Sources

# Chapter 9

# Choosing Lamp Types & Sources

(The High Performance Sustainable Source Equation:
High Luminous Efficacy + High Color Quality +
Positive Life Cycle Assessment =
High Performance Sustainable Lighting Sources)

## DESIGN FACTORS

Lighting fixtures are made up of many components—lamps, reflectors, lenses, baffles, louvers—that are designed to work together to project the light in a particular way: sometimes in a very tightly defined beam to highlight or accent an object or feature, and sometimes in a diffuse pattern to create an ambient light that fills a space. Arguably, the most important components are the light bulbs or *lamps,* as we will refer to them. These are at the heart of our lighting fixtures, and are the actual source of the light. When we think about the type of lighting fixtures to specify for a space, the type of lamp, or the source, is one of the first things to consider. Different sources have different qualities and features that make them suitable for different purposes. Some are more easily controlled, or dimmed, than others. Some are available in a variety of color temperatures. Some are suitable for luminaires designed to provide ambient lighting solutions, and some are suitable for accent lighting. Some come to full illumination instantly, and others have a startup cycle that can take as long as a few minutes. Some can be switched off and on frequently, without any delay, and others need to be allowed to cool down after being switched off before they can be started again. And they all have different efficacies, which is another way of saying they each use different amounts of energy to create the same amount of light. Finally, they each have different lifetimes, and require more or less frequent maintenance and replacement.

## LUMINOUS EFFICACY

The term luminous efficacy as it applies to electric lighting sources refers to the amount of light energy generated per unit of electrical energy used. We usually express the luminous efficacy of a particular electric lighting source as a ratio of lumens to watts. A lumen, as you will recall from Chapter 2, is a unit of energy in the form of visible light. A watt is a unit of energy in the form of electrical current. So the higher the ratio of lumens to watts, the more efficient the source is at converting electrical energy into visible light. One of the first things to look for in a sustainable lighting source is the ability to create a lot of light with a little bit of electrical energy, or a high ratio of lumens produced to wattage used. But as we will discuss later in this section, the luminous efficacy of a source is not, by itself, a good indicator of the amount of light, or the effective lumens, a luminaire incorporating that source will actually deliver to the target area. The natural distribution of light produced by the source, and the efficiency of the luminaire in harnessing that light and projecting it to the target area, are both contributing factors. To complement luminous efficacy ratings for sources, a new metric that has been developed in the last few years is the TER, or *Target Efficacy Rating* (NEMA LE 6-2009), which measures the ratio of lumens emitted from a luminaire that contribute to the illumination of a target area per watt of power used.

## COLOR QUALITY

As we've already discussed, energy efficiency is only part of the story. High performing buildings must also provide healthy, productive environments that satisfy the human requirements of the building's occupants. The same must be true, then, for high performing lighting. The quality of the interior light must be high, and it must satisfy the visual requirements of the occupants. A very important quality of white light sources is their ability to render colors accurately. We've all been in an environment where the lighting, be it poor quality fluorescent or low color-rendering metal halide, makes us look pale or even green. And many people have purchased something only to find that its color looks different at home than it did in the store. These are the results of a poor color quality light. Not so long ago, the color quality of our light was considered important only in high-end retail, hospitality and maybe some front-office corporate environments. We now accept that good color quality is important in many more situations, especially healthcare, as well as all office and retail environments, and even some exterior lighting.

As described in Chapter 5, we measure the ability of electric lighting sources to accurately render the colors of objects by rating the source with a numerical scale called CRI (color rendering index). Sources are evaluated and assigned a CRI rating of 1-100. Daylight, incandescent, and halogen light sources are considered to have a CRI of 100. In fact, the CRI scale was not developed until after the advent of fluorescent and HID lighting created the need by introducing electric light sources that did a poor job of rendering colors. Though incandescent and halogen may be considered perfect, high color rendering can be achieved with fluorescent, metal halide, and LED sources. There is some controversy regarding the use of CRI as a reliable metric, especially for LEDs. A new metric, the Color Quality Scale, under development by researchers at the National Institute of Standards and Technology, is considered by some to be a better gauge, especially when evaluating LED lighting. But for the time being, the CRI scale is the tool available for evaluating light sources. Any good lamp or LED chip manufacturer will publish the CRI of their products, and in general lamps and sources with a CRI rating of at least 80 or 85 should be specified wherever accurate color rendering is important. LEDs tend to do a better job rendering saturated colors, even with lower CRI ratings, and so with LED fixtures we should review an actual sample and leave it to our eyes to judge whether it produces light of a high enough color quality for the application at hand. But for fluorescent and high intensity discharge lamps, we should always specify lamps with as high a CRI as we can, and limit our specifications to those that score no lower than 80 and preferably 85 and higher.

## LIFE CYCLE ASSESSMENT (LCA)

Another, very important factor when considering the quality and sustainability of a lighting source is the impact it has on the environment throughout its life cycle. When looking at the life cycle assessment of a product, considerations include: how much energy is expended to produce the product; how much waste is created as a result; the energy costs of bringing the product to market; what natural materials need to be gathered and mined for the product's manufacture; what, if any, dangerous materials are contained within the product; and what happens to these materials when the useful life of the product is over. So, when we consider a light source's sustainability, along with questions about efficacy, energy efficiency, and quality, we need to ask what's in our lighting sources, how long will they last, and where are they going to end up. If a light source, or lamp, is energy-efficient, but is made with mercury or other toxic substances that need to be segregated from the environment at the end of its life, that needs to be considered. If these toxic materials can be safely recycled and reused, that's one thing. If they are difficult to extract, recycle and reuse, and are thus more likely to be incorrectly handled and allowed to reenter the environment, that's another.

It's important to look at the big picture when adding up the sustainability of today's light sources, and unfortunately, as

| Source Type | Approx. Efficacy (lm/watt) | Lamp Life (Hours) | CRI | Dimmable | Toxic Materials? |
|---|---|---|---|---|---|
| Incandescent | 10-17 | 750-6000 | 100 | Yes | No |
| Tungsten Halogen | 16-26 | 2000-6000 | 100 | Yes | No |
| 4' T12 Fluorescent | Up to 82 | 24,000*** | 60-80 | Yes | Yes |
| 4' T8 Fluorescent | Up to 95 | 24,000-55,000**** | 80-85 | Yes | Yes |
| High Performance T8 | Up to 102 | 24,000-55,000**** | 80-85 | | Yes |
| 4' T5 Fluorescent | Up to 96 | 30,000-40,000**** | 85 | Yes | Yes |
| 4' High Output T5 Fluorescent | Up to 89 | 30,000-60,000**** | | | Yes |
| Compact Fluorescent (not self-ballasted) | Up to 70 | 10-20,000 | 82 | Yes | Yes |
| Magnetic Induction | Up to 84 | 60- 85,000* | 80+ | 50% | Yes (Solid Form) |
| Metal Halide & Ceramic Metal Halide | Up to 110 | 10-20,000 | 60-90+ | 50%** | Yes |
| Sodium Vapor | Up to 105 | 24,000 | 0-20 | 50%** | Yes |
| LED Downlights (Complete Fixtures!) | Up to 60+ | 50-60,000* | Up to 90+ | Yes | Very Small Amounts In Solid Form |
| LED Troffers & Area Lighting (Complete Fixtures!) | Up to 100+ | 50-60,000* | Up to 90+ | Yes | Very Small Amounts In Solid Form |

**Figure 9-1. Efficacies, CRI, and Lifetime for Various Lamp Types:** All values are approximate and will vary for individual lamps & manufacturers. Fluorescent and HID efficacies are approximate projections that take into account best-case ballast/lamp systems. LED efficacies are based on absolute photometry measurements for complete luminaires and are not subject to correction for fixture inefficiencies. *Lifetime to 70% Lumen Depreciation **With Color Shift ***3hr. start **** 12 hr., program start

the table in Figure 9-1 shows, these equations do not always lead to a clear conclusion. While fluorescent, and to a lesser extent metal halide sources, may contain hazardous materials in gaseous form, the benefit of using them, with their high efficacy, may outweigh the risk and cost of disposal, at least in the short run. Magnetic induction lighting seemed like a promising technology 10 or 15 years ago, though recent improvements in ceramic metal halide and LED lighting have outstripped it. It seems clear that LED sources hold the greatest promise. With a relatively high and ever increasing efficacy, potentially high color quality, long lifetime, and the potential to contain very minimal amounts of toxic materials, they are shaping up to play a major role in the future of electric lighting. But we have more to learn regarding the sustainability of LED lighting with regard to the manufacture of LED nodes, including the impact of mining, transporting and processing the materials that go into them, and the various toxic materials used.

# Chapter 10

# Lamps, Source Types, & Relative Photometry

## INCANDESCENT LAMPS

With a few minor improvements along the way, the modern incandescent lamp still functions under the same principles as when Thomas Edison ran the first successful test, in 1879, in which a lamp remained lit for 40 hours before burning out. The incandescent lamp works via a simple mechanism. Electricity is run through a filament composed of a material with the requisite electrical resistance to cause it to heat up and glow, much the way a heating element will glow in a toaster. The filament is held within an evacuated glass envelope or bulb, or one containing inert gases in place of air, that keeps the filament from burning out too quickly. Incandescent lamp technology is a very inefficient method of creating light from electricity. In addition to having a low efficacy (only about 10-17 lumens are generated per watt of electricity), about 90% of the energy used is converted into an unintended by-product: heat. So incandescent lamps not only use a lot of energy to generate a small amount of light, but also create an additional load for air conditioning systems, resulting in increased cooling costs and energy use. Many countries, including the United States, have introduced legislation to regulate, or to outright ban, the manufacture and sale of the most inefficient incandescent lamps. In the United States this legislation, the Energy Independence and Security Act of 2007, takes the form of regulations that set different efficacy standards for incandescent lamps of different wattages. It is generally agreed that we need to find an energy-efficient replacement for household, incandescent lamps. But much of the public at large is resistant to this phase-out, as these lamps are cheap to purchase, easy to dim, are not made with toxic materials, and create a very pleasing quality of light.

**Life Cycle**

Incandescent lamps do not pose any challenges for disposal, but they are also short-lived, typically lasting between 750 and 6,000 hours, with some types having a significantly shorter lifespan.

**Photometric Properties**

The glowing filaments in incandescent lamps produce an omnidirectional distribution of light, which means they project light in all directions. Energy efficiency aside, this makes them especially well suited to applications that incorporate large diffusers and decorative, translucent shades to produce larger sources of soft, ambient light. Many of our traditional floor and table lamp designs have grown out of the widespread use of this lamp type and its ubiquity in our homes for so many decades.

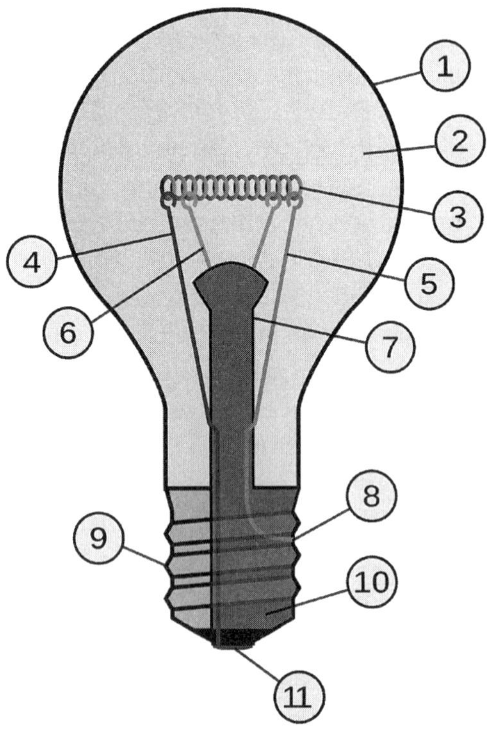


**Figure 10-1. Diagram showing the major parts of a modern incandescent light bulb: 1. Glass bulb, 2. Inert gas, 3. Tungsten filament, 4. Contact wire (goes to foot), 5. Contact wire (goes to base), 6. Support wires, 7. Glass mount/support, 8. Base contact wire, 9. Screw threads, 10. Insulation, 11. Electrical foot contact**

The relatively large size and omnidirectional nature of the incandescent lamp makes it less well suited for applications requiring a focused beam, or any kind of well-defined projection of light, though some directional, incandescent reflector lamps are available. Incandescent lamps score a perfect 100 on the CRI scale, and they produce a very warm color temperature of light, about 2,700K. They are traditionally used in homes, retail stores, and hospitality environments.

## HALOGEN LAMPS

Halogen lamps are very similar to incandescent lamps in their general workings. A filament is energized with an electrical current that causes it to heat up and glow. The main difference is that the filament of a halogen lamp is made of tungsten (a type of metal), and is enclosed in a smaller envelope filled with an inert gas and a halogen. (Halogens are a group of non-metal elements.) The combination of the halogen and tungsten allows the lamp to be run at a higher temperature without burning up. The result is a slightly cooler color of light and a slightly higher luminous efficacy (up to a maximum of 26 lumens per watt), making these lamps marginally more energy-efficient than traditional incandescent lamps. Like incandescent lamps, halogen lamps are dimmable through the full range of their output.

### Life Cycle

Like their cousins the incandescent lamps, halogen lamps do not pose any challenges for disposal, but they are also short-lived, typically lasting between 2,000 and 6,000 hours, with some types having a significantly shorter lifespan.

### Photometric Properties

Halogen lamps, like incandescent lamps, are based on a filament that emits an omnidirectional distribution of light, though the potentially small size of their filaments make them more easily incorporated into precise reflector lamp designs that integrate an optical system (a reflector and lens) to produce a very tightly

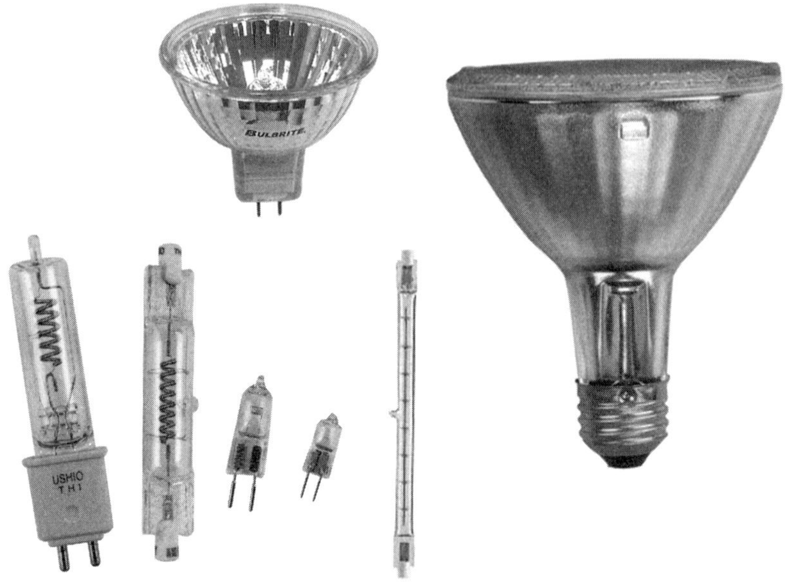


**Figure 10-2. Halogen lamps, with their potentially small filaments and envelopes come in many shapes and sizes.**

controlled beam. The potential for smaller filaments and envelopes has the added advantage of enabling the design of many different low profile luminaires. All these qualities make halogen lamps a very good choice for accent lights, display lights, and theatrical spotlights, though the larger halogen lamps, when used with the right kind of reflector and luminaire housing, also make good floodlights and indirect lighting treatments. Halogen sources produce light with a color temperature of 3,200K, a little cooler than incandescent lamps but still warm and inviting, and have been traditionally used in homes, retail, and hospitality environments. *Halogen infrared* (HIR) lamps are the most efficacious in this category as they incorporate a reflective coating on the filament tube of the lamp that re-directs the emitted infrared energy (heat) back onto the filament. This increases the lumen output for the same amount of energy used when compared to standard halogen lamps. Many of the most commonly used halogen lamps operate with low voltage current (12 or 24 volts) and require a transformer, usually incorporated in the body of the luminaire.

## METAL HALIDE LAMPS

Metal halide and ceramic metal halide (CMH or CDM) lamps fall under the category of high intensity discharge (HID) lamps. HID lamps produce light by passing an electric arc through a pressurized mixture of gases and metal salts in a transparent capsule or tube. In a metal halide lamp, the arc is passed through a small tube containing a high-pressure mixture of argon, mercury, and a variety of metal halides. The exact makeup of this mixture affects the quality of light produced, including the color temperature and luminous intensity. Metal Halide lamps are another high efficacy source, producing up to 110 lumens per watt of electricity (after allowing for ballast related losses), making them a very sustainable option. Larger metal halide lamps are widely used in outdoor lighting as well as in many industrial and commercial settings. In recent years, smaller, lower wattage, ceramic metal halide lamps and luminaires have been put to good use as replacements for halogen fixtures in retail and high-end commercial applications. Because these lamps can be quite small, they are easily integrated into low-profile luminaires, incorporating a variety of reflectors and lenses to produce a very controllable, high quality light. Metal halide lamps are only dimmable to about 50% of their full output. And dimming them can produce a color-shift in their light output. In an industrial setting it is possible to incorporate step-dimming ballasts with metal halide lamps as part of an energy efficiency controls solution, but for most other application we cannot consider metal halide when a dimmable source is required. Metal halide lamps operate under relatively high temperatures, and require a warm-up period of over a minute or two during which time the light goes through a significant shift in color. Once extinguished, they cannot be restarted until they have cooled off, making their use problematic

**Figure 10-3. Metal Halide and Ceramic metal Halide Lamps come in many shapes and sizes.**

in scenarios where the lighting needs to be switched on and off while the occupants are present, or for emergency lighting.

**Life Cycle**

Metal halide lamps have a significantly longer lifespan than incandescent or halogen lamps, which helps reduce waste and contain maintenance costs. But they are made with small amounts of mercury that is damaging to the environment if not disposed of properly.

**Photometric Properties**

Metal halide lamps are available in a wide variety of color temperatures and CRI ratings. As they produce a point source of light, like incandescent and halogen, metal halide lamps can be designed for a very general, omnidirectional, wide distribution of light, and they can also be successfully wrapped in small envelopes, and integrated with reflectors and lenses, to create highly controllable and precise accent lighting.

Metal halide lamps (and all HID sources) require a ballast, a piece of electrical equipment that regulates the current to start and run the lamp, to function properly. These ballasts are usually integrated into the body of the luminaire.

## SODIUM VAPOR LAMPS

High pressure sodium and low pressure sodium vapor lamps are HID sources, like metal halide, that have traditionally been used in industrial and exterior lighting applications. High pressure sodium lamps have a very high luminous efficacy (over 100 lumens per watt), and low pressure sodium lamps even higher (nearly 200 lumens per watt), but they both emit a yellowish light with poor color-quality and a color rendering index of 20, at best. (Low pressure sodium light is monochromatic, with a CRI of 0.) They are considered a legacy technology and are being gradually replaced by metal halide and, more recently, LED sources. As with the other HID sources, they require a warm-up period of over a minute during which time the light goes through a significant shift in color. Once extinguished, they cannot be restarted until they have cooled off. In the context of high-performance design they are obsolete, and can only be recommended for applications in which it is not at all important to be able to discern differences in the colors of objects—applications that are few and far between—or where a yellow colored light is desirable for a specific effect.

## FLUORESCENT LAMPS

Fluorescent lamps first became available in the late 1930's as an energy-efficient electric lighting source for use in the industrial and commercial sectors. The fluorescent lamp generates light in an entirely different manner than either incandescent or halogen lamps. It is a more complicated system and it requires the use of a ballast, which is an electrical device that regulates the current to first start and then run the lamp. The fluorescent lamp is a form of gas-discharge tube that uses electricity to excite a mercury vapor inside the lamp envelope, producing an ultraviolet light in the process that in turn excites a white phosphorescent coating on the inside surface of the lamp. The phosphorescent material glows brightly, emitting enough light to be useful as a lighting source in a luminaire. Fluorescent lamps have a far higher luminous efficacy than incandescent or halogen sources, with the more efficacious types producing over 100 lumens per watt of electricity (after allowing for ballast-related losses). And fluorescent lamps are dimmable, with some types able to dim down to as little as 1% of their full output. Dimming of fluorescent lamps requires the use of a dimming ballast, the specification of which will determine the degree to which the lamps can be successfully dimmed before flickering or being extinguished.

**Life Cycle**

Fluorescent lamps have a significantly longer life than incandescent or halogen lamps, which helps reduce the generation of waste and lowers maintenance costs. This lifespan can vary depending on the ballast type used to run the lamp, as well as the frequency with which the lamp is switched on and off. The more often the lamp is switched, the shorter its lifespan will be. One drawback of fluorescent lamps is that, like high intensity discharge lamps, they are made with small amounts of mercury which damage the environment if not disposed of properly. Care should always be taken when handling these lamps to avoid accidental breakage, and they should only be disposed of through an appropriate and approved waste recovery system. Some municipalities now have programs to facilitate the collection and disposal of these lamps along with other toxic materials.

**Photometric Properties**

The quality of fluorescent lighting has a poor reputation among many consumers, mainly due to bad experiences with lower quality lamps and poorly maintained systems. However, high color-rendering fluorescent lamps are available today in a variety of warm and cool color temperatures, and can be suitable for high-end retail, residential, and hospitality installations. Most fluorescent lamps are relatively large for the amount of light they produce when compared to a point source like a halogen or metal halide lamp. They have a greater luminous surface area, and they project an omnidirectional distribution of light, making them a very good choice in applications where soft, diffused, ambient lighting effects are desired. Conversely, they are harder to control with reflectors and lenses, and are less well suited for use in accent lighting, and where a narrow distribution of light over long distances is required. Fluorescent sources are also negatively affected by cold ambient temperatures, and will not generally perform to their full potential in exterior applications in colder climates, though there are some lamp and ballast combinations that can be used down to -20 degrees Fahrenheit in enclosed fixtures.

**Ballast Factor & Ballast Efficiency Factor**

As with other discharge lamps, the amount of power required to drive a fluorescent lamp is dependent on the ballast used. Different ballasts can drive a fluorescent lamp to a higher or lower degree, and will require more or less power to do so. As a result the input wattage of a two-lamp fixture, where each lamp is rated at 28 watts, can actually vary depending on the specific ballast that is used, and is often more than the combined lamp rating of 56 watts. Similarly, the lumen output of the lamps will fluctuate depending on the ballast used to drive them. To denote this, fluorescent ballasts will carry a published *ballast fac-*

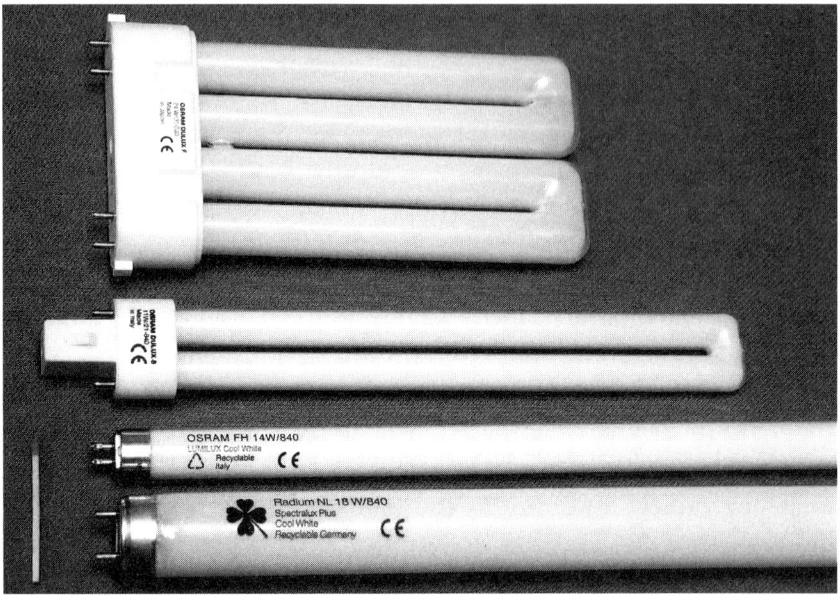


**Figure 10-4. Linear fluorescent T8, T5, and compact fluorescent lamps. Photo Credit: Christian Taube, [CC-BY-SA 2.0 (www.creativecommons.org/licenses/by-sa/2.0)], via Wikimedia Commons**

*tor* as part of their specification. Ballast factor is defined as the light output, or lumens, a ballast will produce from the lamps it is designed to run when compared with a reference ballast that operates the lamps at their specified nominal power rating. A ballast with a ballast factor of 1.0 will run the lamps at their nominal power rating, and produce those lamps' nominal rated lumens. A ballast factor of 0.9 will produce 90% of the lamps' rated lumens, and a ballast factor of 1.1 will produce 110% of the lamps' rated lumens. On the one hand this may seem to complicate the question of how much light is produced by fluorescent sources and how much electricity is required to drive them, but on the other, this relationship between fluorescent lamps and ballasts can be a valuable tool for the lighting designer. It allows us to specify one ballast in a situation where we need less light, and want to use less energy, and another where we need more light from the same lamps. Interestingly enough, the latter situation can sometimes be a strategy for energy cost-effectiveness and energy efficiency. Enhanced lumen output from lamps, even at a greater energy cost, can sometimes allow us to use fewer fixtures, or fixtures with fewer lamps, resulting in overall energy savings. Along with ballast factor, there are also losses and inefficiencies inherent in the operation of all ballasts that must be factored in. So a fluorescent ballast with a ballast factor of 1.0 will still have an input wattage that is greater than the sum of the lamps' nominal power rating. In the cases of ballasts with very low ballast factors, the input wattage of the lamp/ballast system can be less than the nominal power rating of the lamps. The relationship between the input wattage and the ballast factor is defined as the *ballast efficiency factor*. The ballast efficiency factor is the result when the ballast factor is multiplied by 100, and then divided by the input wattage of the ballast. The higher the number (when compared with the same type of ballast, running the same number of lamps) the more efficient that ballast is.

Formula to arrive at Ballast Efficiency Factor (BEF)

(BEF) = [Ballast Factor (BF) X 100]/Ballast Input Wattage

| Description | Number of Lamps | Length | Input Watts | Ballast Factor |
|---|---|---|---|---|
| T5 28 W | 2 | 4 | 64 | 1 |
| T5HO 24 W | 2 | 2 | 52 | 1 |
| T5HO 54 W | 2 | 4 | 120 | 1 |
| T5HO 54 W | 4 | 4 | 240 | 1 |
| T8 25 W | 2 | 3 | 48 | .91 |
| T8 25 W | 2 | 3 | 48 | .98 |
| T8 32 W | 2 | 4 | 59 | .88 |
| T8 32 W | 2 | 4 | 59 | .87 |
| T8 32 W | 3 | 4 | 90 | .97 |

**Figure 10-5. Partial specification chart for a variety of fluorescent ballasts, showing input wattage and ballast factor. These values will be different for each manufacturer and ballast type.**

**Formats & Efficacies**

Fluorescent lamps come in many shapes and sizes: linear, circular, and spiral, to name a few. And there are many more configurations of compact fluorescent lamps, which are often comprised of multiple linear tubes folded back on themselves. Linear lamps are by far the most prevalent fluorescent lamp type, and the trend over the years has been toward lamps with a smaller and smaller diameter, and with greater and greater efficacy. The T12 lamp, so named because it is a tube with a diameter of twelve-eighths of an inch, has been largely supplanted by the T8 and T5, both smaller profile lamps with greater luminous efficacies. But obsolete systems of T12's run with old, magnetic ballasts are still ubiquitous, providing some of the easiest to implement, lowest hanging fruit of today's energy-efficient retrofit market. This outdated lamp-ballast combination is significantly less efficacious, and therefore less energy-efficient, than T8 or T5 lamps run by newer, electronic ballasts. In addition, many of the older magnetic ballasts contain concentrations of PCB's, a toxic material whose manufacture and sale was banned in the United States in 1979. Countless numbers of these T12 systems are still in use, though these old lamps and ballasts can be easily replaced in existing lighting fixtures. The savings in energy costs, along with government and utility incentives available to assist

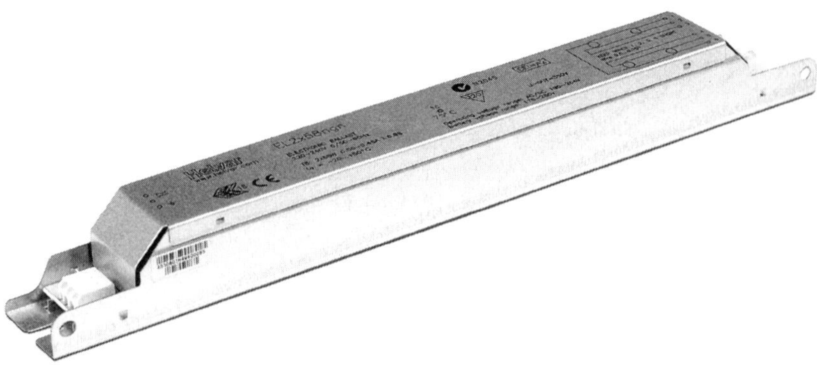

**Figure 10-6. Electronic fluorescent ballast. Photo Credit: Symppis, [CC-BY-SA 3.0 (www.creativecommons.org/licenses/by-sa/3.0)], via Wikimedia Commons**

owners who want to do these retrofits, will often go a long way towards paying for the cost to purchase and install new lamps and ballasts. Thanks to government regulation, most magnetic T12 ballasts are now illegal to manufacture and sell in the United States, just as many of our common incandescent lamps will become unavailable starting in 2012.

**High Performance T8 Systems**

High performance T8 (or HPT8) systems constitute one of the most successful strategies for energy efficiency with fluorescent lighting. An HPT8 system combines a high lumen, long-life, T8 fluorescent lamp with a high-efficiency electronic ballast. HPT8 lamps are defined as four-foot, 32, 28 or 25 watt, T8 lamps, with a minimum initial lumen output, mean lumen output, CRI, and lifetime rating. HPT8 ballasts are defined as electronic ballasts that meet or exceed a minimum ballast efficiency factor (BEF). The minimum BEF is different for different types of ballasts, depending on the number of lamps the ballast is designed to run and whether it is an instant-start ballast or a program-start ballast. (Instant start ballasts are more efficient, but program start ballasts are recommended for use with occupancy sensors, or where frequent switching is expected, as we will discuss in greater length in Chapter 13, which covers controls.) In addition, HPT8 ballasts are available in low, normal, and high ballast factors. By combining a high-lumen lamp with a low ballast factor HPT8 ballast, the same amount of light one would normally get from a standard output lamp and ballast combination can be achieved with a lower input wattage. HPT8 systems can also be specified with normal or high ballast factor ballasts, and in this case the increased lumen output may allow the designer to specify fixtures with fewer lamps, or simply reduce the number of fixtures in a layout, resulting in an even greater savings in energy cost for the same amount, or even increased, light output. Extra long-life lamps can also be specified, but the lumen output will be reduced with these. Detailed specifications, as well as a list of qualifying HPT8 lamps and ballasts, can be found on The

Consortium for Energy Efficiency (CEE) website: http://www.cee1.org/com/com-lt/com-lt-main.php3

**Self-ballasted Compact Fluorescent Lamps**

A popular choice today for consumers and businesses who want to save energy without having to purchase and install new lighting fixtures, the self-ballasted compact fluorescent lamp combines a lamp and ballast in one device and can replace incandescent lamps in existing fixtures. This will always be less expensive than installing a new fixture with an integral ballast. While it may be an effective short-term energy-efficiency solution for some pre-existing lighting systems, this technology has drawbacks. The lamp and ballast combination will often result in a profile that does not fit well in an existing fixture, and one that makes poor use of the reflectors and baffles therein that are usually designed to optimize the distribution of light from an incandescent lamp. And the quality of these self-ballasted lamps can vary. They seldom come with published specifications, and the integral ballasts may perform more or less well. They may also have a low power factor, which can create a greater burden on the electrical utility power plant, and as a result they may use more energy than an incandescent lamp of equal wattage. And, like conventional fluorescent lamps, they contain toxic waste that needs to be disposed of carefully. Unfortunately, especially in the consumer market, these are all too commonly thrown in the trash and taken to the local landfill where the mercury within can leach into the surrounding environment.

## MAGNETIC INDUCTION LAMPS

By no means a new technology, the induction lamp, which was first developed by Nicholas Tesla in the 1890's as a competing alternative to Thomas Edison's incandescent lamp, has lately experienced a rebirth in the lighting industry. These lamps are variants of the fluorescent family in that they emit light via a

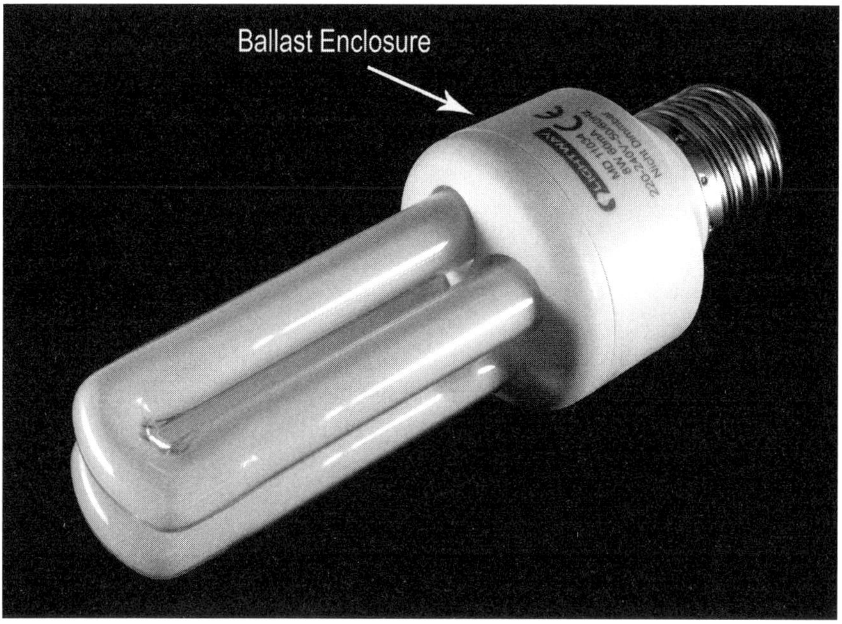


**Figure 10-7. Self-ballasted compact fluorescent lamp. Photo Credit: Armin Kübelbeck, [CC-BY-SA 3.0 (www.creativecommons.org/licenses/by-sa/3.0)], via Wikimedia Commons**

phosphorescent coating on the inside of a sealed glass envelope. Induction lamps are energized via a process whereby electromagnets wrapped around a tube-shaped lamp, or held inside a cavity in a bulb-shaped lamp, produce a strong magnetic field within the glass envelope. The magnetic field excites mercury atoms, provided by a solid pellet of mercury amalgam, to produce an ultraviolet light. The ultraviolet light in turn excites the phosphorescent material, causing it to glow brightly enough to be useful as a lighting source in a luminaire. Induction lamps are very long-lived, and in tandem with the electronic ballasts that run them can produce as many as 84 lumens per watt of electricity. They are dimmable to some degree, in some systems down to about 50% of the lamps full output, but not nearly as effectively as standard fluorescent lamps are. As a result they are not suited to applications where low level dimming is required.

[11] Induction lighting is considered by some to be a promising technology for high-performance applications, but the steady increase in efficacy of both HID and LED sources has already outstripped it. Thanks to their long life induction lamps can still be an appropriate choice in some cases where conditions make the lighting systems difficult to maintain—for the time being, anyway.

**Life Cycle**

Induction lamps last approximately four times as long as their standard fluorescent counterparts. And the lamp's lifespan is not negatively affected by frequent switching, making these a

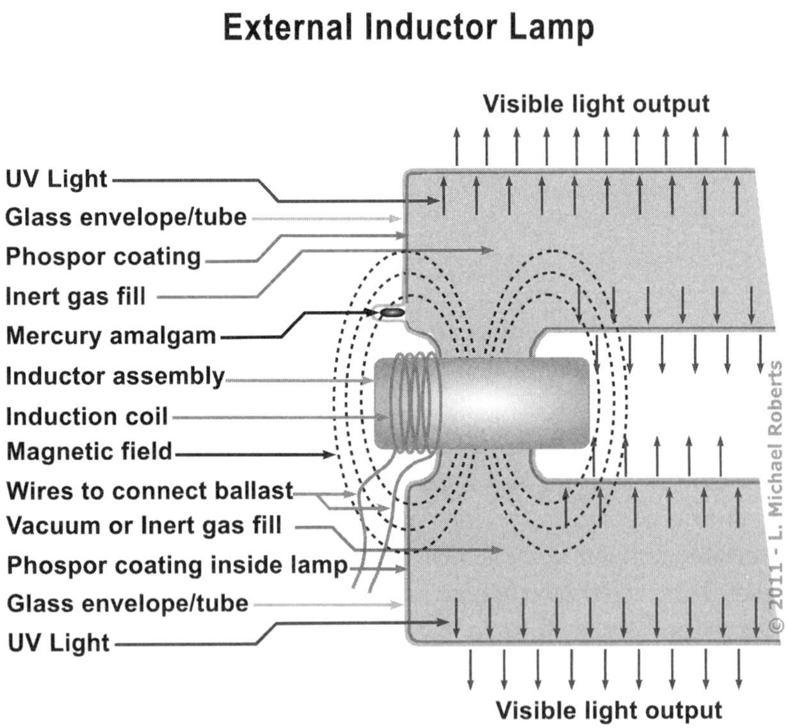


**Figure 10-8. Diagram showing the process by which light is generated with a magnetic induction lamp utilizing an external inductor. Image credit: L. Michael Roberts, Indulux Technologies, Inc.**

good choice where occupancy sensors will be employed (as we will discuss in a later chapter). In addition, since the mercury in these lamps is captured in a solid form (mercury amalgam), the toxic component is easily recovered from spent lamps for proper disposal or reuse. The biggest physical difference between induction lamps and regular fluorescent lamps, and the reason induction lamps are so long-lived, is that they are electrodeless. All the other lamps we have discussed depend on electrodes: metal conductors that bring electrical current to the interior of the lamp envelope to energize the lamp. The electrodes are the physical weak link, subject to breakage from shock and vibration, and to depletion over time as their material is stripped away by the electrical current. In addition, the incorporation of electrodes inside a lamp envelope necessitates having a break in the glass where the electrodes can be introduced, which is then resealed, as is the case with the ends of a fluorescent tube or the base of an incandescent light bulb. This seal is another weak link, and is a prime site for breakage or the development of leaks whereby the interior composition of the lamp, whether a void, an inert gas, or a mercury vapor, can be compromised.

**Photometric Properties**

As the method whereby light is generated is the same for both lamp types, the photometric properties of magnetic induction lamps are virtually the same as those for fluorescent lamps. Since their light is the product of a phosphorescent coating, which can be tailored to emit different spectra of visible light, they are available in a variety of warm and cool color temperatures. Induction lamps come in a variety of shapes that make them a suitable source for recessed downlights and industrial lighting fixtures. And they are far less sensitive to colder temperatures than the conventional fluorescent sources, making them a good choice for exterior applications in climates where fluorescents might not perform as well. But their large size makes the light they produce difficult to control, reducing the effective lumens an induction fixture will deliver to its intended target. As the ef-

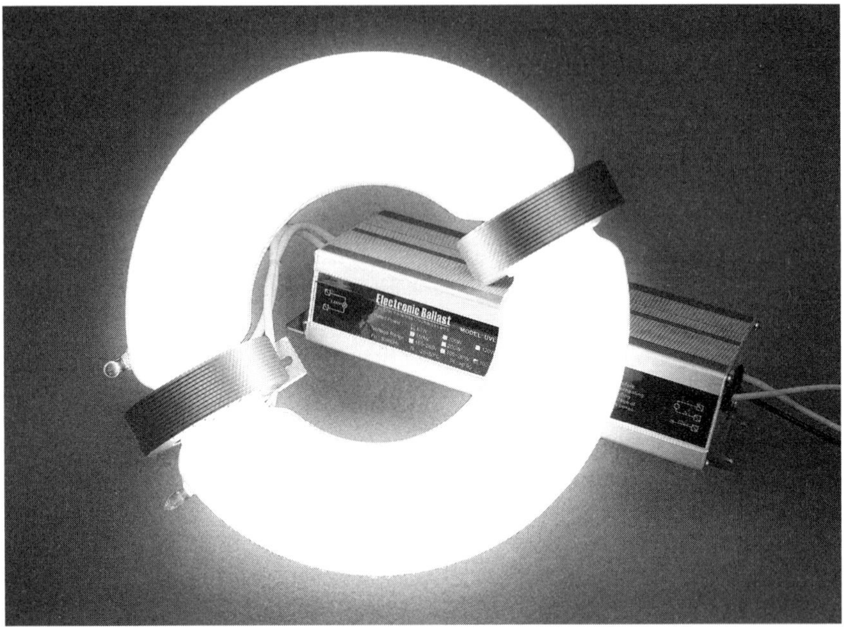


**Figure 10-9. Magnetic induction lamp with external electromagnets wrapped around a circular lamp envelope. Photo credit: Indulux Technologies, Inc.**

ficacy of LED sources continues to rise, and as the price for this most promising technology drops, the likelihood that magnetic induction will ever play a major role in high performance lighting will continue to decrease.

## LIGHT EMITTING DIODES

LED lighting, (also known as SSL, or solid state lighting) is a very exciting technology beginning to come into its own in the lighting industry. LEDs (light emitting diodes) create light by energizing very small semi-conductor chips to cause an electroluminescent reaction that makes them glow brightly. LED light engines—the circuit boards and arrays of LEDs that make up the "source" in an LED luminaire—can be made with minimal toxic materials, all of which are encapsulated in solid forms and remain easy to extract and dispose of properly. When energized, LEDs create directional, point sources that are quite small, making the light they emit easy to direct and control with lenses and reflectors, opening up a wide range of luminaire design possibilities. They are dimmable, switch on instantly, last a long time, and their luminous efficacy is potentially very high, making this technology a logical replacement for many of our current conventional sources. But with these benefits come some technical challenges and, for the time being anyway, increased costs. Today's LED luminaires are a practical, cost-effective solution for some applications, and the promise of more cost-effective highly efficient LED luminaires in the near future is bright. But there are many ways in which this LED revolution demands a new way of thinking about sources and luminaires. The following section provides an introduction to the concept of relative photometry, which we use to calculate the delivery of light from luminaires employing conventional sources. In the next chapter we will discuss the concept of absolute photometry, and the reasons why LED sources, and their respective luminous efficacies, have to be considered according to this method of calculation.

## SOURCES, APPLICATIONS & FIXTURE EFFICIENCY

When we look at the energy efficiency of our traditional (non LED) lighting, we may start with the lamp, or light source, and consider the efficacy or lumens per watt we can expect to generate with that particular lamp. But we also have to look at the efficiency of the lighting fixture in which we are employing that lamp. Different lighting fixtures may do a better or worse job of utilizing the light generated, and this can greatly reduce the effective lumens that are ultimately delivered by the fixture. Fixtures with well-designed reflectors and lenses will obviously do better than those where optical engineering was not adequately considered. But the efficiency of the fixture is not only a product of the quality of the fixture. It is also a product of the suitability of the lamp, or source, to the treatment that fixture provides. Simply put, any fixture that makes more effective use of the natural distribution of light from the source it employs will be more efficient than one in which more reflectors, louvers, and baffles are required to direct, focus, or shape the light. And since we often need to direct, focus, and shape the light from our conventional sources, in many instances we will require the use of fixture types that necessarily result in a less efficient use of that source's output.

## RELATIVE PHOTOMETRY

We use relative photometry to express this relationship between lamps and luminaires. With relative photometry, the lumen output of the lamp and the efficiency of the fixture are both used to calculate the effective lumens, or the amount of light actually projected by the fixture after allowing for the losses created by any light-shaping components that fixture may employ. For example, a relatively open direct/indirect luminaire may

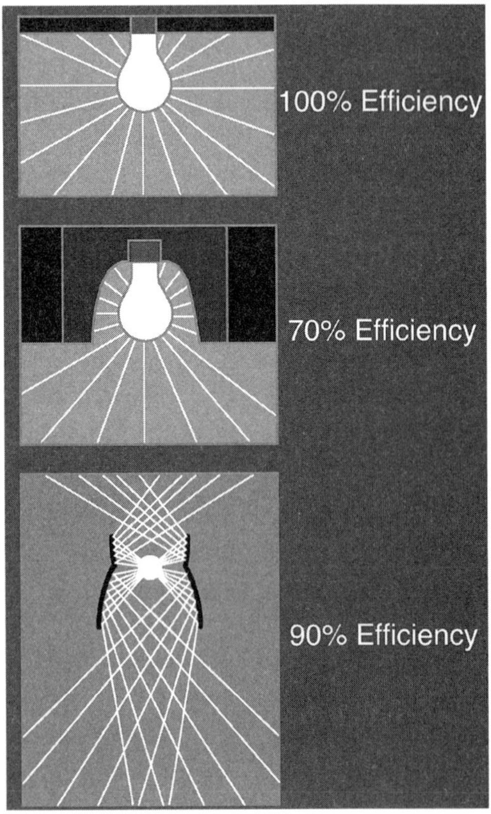


**Figure 10-10. Recessing a lamp into a luminaire housing and employing optical modifiers like reflectors, baffles and lenses will necessarily result in the absorption of light and the lessening of the fixture's efficiency.**

make very good use of a fluorescent lamp's natural distribution—a nearly three hundred and sixty degree projection of light. But a recessed down light will necessarily require reflectors and baffles to direct all of that same lamp's output into the space below the ceiling plane, resulting in a significant degree of inefficiency in the process. As illustrated in Figure 10-10, the use of a reflector, or lens, to shape and direct the light, or the employment of baffles and shields as glare protection, no matter how well engineered these components may be, will result in the absorption of light energy and reduced fixture efficiency. This is unavoidable.

Most of our conventional sources produce an omnidirectional distribution of light, making the use of all of these beam-shaping devices necessary in many luminaire types. LED sources are directional by nature, making it easier to deliver a greater number of the lumens they generate to a specific target. So LED luminaires are inherently better at delivering effective lumens. But there are some important differences between LEDs and conventional sources that make it impossible to use relative photometry to calculate the amount of light that will be delivered by an LED luminaire. We will discuss these differences in detail in the following chapters.

# Chapter 11

# LED (SSL) Lighting

## WHAT IS AN LED?

*LED,* which stands for light emitting diode, and *SSL,* which stands for solid state lighting, are two different names for the same technology currently revolutionizing the lighting industry. Like fluorescent and HID sources, LEDs require the integration of an electrical device into the luminaire, or the lighting system, to regulate the current that energizes the diode to make it glow. With fluorescent and HID sources this device is called a ballast, and in the case of LEDs it is called a driver. LED sources have the potential for very long life and very high luminous efficacies. They come on instantly, with no warm-up time or flicker. They are dimmable. They are small point sources that can be easily incorporated with optical systems comprised of reflectors and lenses to produce a highly controllable distribution of light, allowing the luminaire designer to create fixtures that will precisely deliver the light in whatever way the application requires. And they can produce light of different colors from across the visible spectrum, as well as white light in a variety of color temperatures. While white light LED sources of cooler color temperatures tend to have higher luminous efficacies than the warmer ones, today's warm white LED luminaires will still beat the energy efficiency of incandescent and halogen equivalents by a good margin: generating the same amount of light using only one-fifth to one-sixth the energy. They are at least twice as efficient as most compact fluorescent fixtures. But don't pay too much attention to these comparisons, because by the time this book is printed the performance of the currently available LED fixtures will be even better. Clearly, this technology has great promise, and as the cost to produce good, reliable LED fixtures drops, it is likely to supplant many of our conventional lighting sources in many applications. But LEDs are fundamentally different than conventional sources, and this technology demands an entirely new way of thinking about lighting fixtures and sources and how they work together.

## COLOR MIXING & WHITE LIGHT

At the heart of a light emitting diode is a small semiconductor chip that glows brightly when energized with an electrical charge. Depending on the semiconductor material used, the LED will glow with a different color. For example, an LED with a semiconductor die composed of indium gallium nitride (InGaN) will glow blue, and one composed of aluminium gallium indium phosphide (AlGaInP) will glow red. By using different materials, or combining these materials together in different ways, LED chip manufacturers can create sources that emit light in a wide variety of colors from across the spectrum—but not white light. So how do we make white light with LEDs? The first method, developed and commercialized in the 1990's, was to mix the colors from an RGB LED luminaire. RGB stands for red, green, and blue, the primary colors of an additive color mixing system which, when mixed together, will produce white. The RGB LED luminaires were made with an array of LEDs, some red, some green, and some blue. By energizing these different colored LEDs, at different intensities, a multitude of colors of light could be created. Theoretically, when they were all on at or near their full level they would mix together to create white light. (RGB is also the same color mixing system that creates a myriad of colors in each pixel, or point, on your television or computer screen.) But the quality of the white light these fixtures produced was less than ideal, as LEDs do not necessarily glow with the pure primary colors required for true additive color mixing. Many of the manufacturers of this type of SSL fixture began to add additional LEDs of different colors to make up for this shortcoming. In fact, one manufacturer employs a color mixing system that relies on the combination of seven different colors to achieve a greater representation across the visible spectrum, allowing the creation of many more colors and a truer white light. In addition, the design of these RGB fixtures was such that each color was generated from a separate diode, or point, and this caused a multi-color fringing effect when a shadow was cast, or off-colored scallops in places where the light did not mix completely. The fringing and colored scalloping was mitigated with the invention of the tri-node LED, a single LED device made up of a red, green, and blue die in one package, but the white light these fixtures generated was still not a true, high quality white.

## APPLICATIONS FOR RGB LEDS

Even with these faults, this style of LED fixture remains especially successful in the context of theatrical lighting where the traditional way to create different colors of light has been to use colored filters in front of very inefficient halogen sources. The filters act to restrict the wavelength of the light that passes through them, essentially absorbing—and wasting—a large percentage of the luminous energy emitted, compounding the inefficiency of this system. In addition, where once a lighting designer had to use three or more systems of lighting fixtures, each with a different colored filter, to create a range of colors, the new LED luminaires allow these three systems of inefficient lighting fixtures to be replaced by one, substantially more efficient system of RGB LEDs. Now designers can achieve the same effect with a third the number of fixtures, and a small fraction of the energy. Another application in which this style of lighting fixture remains popular is for the illumination of special architectural features, where the ability to generate thousands of different colors by varying the brightness of the red, green, and blue diodes can create many different moods and kinetic effects. Though this kind of treatment was once limited to nightclubs, restaurants, and other entertainment venues, we are now seeing it incorporated into all kinds of commercial, public, and even residential spaces. Figure A-4 in Appendix A shows the array of red, green, and blue diodes in an RGB lighting fixture. Such a fixture can be used as a theatrical luminaire, an architectural wall washer, or an architectural uplight, and through different programming sequences can produce a wide variety of multicolored, animated lighting effects. This technology has yet another very popular application which has lately begun to create an overlap between the lighting and video display industries by using individual luminatires as pixels and arraying them to create a media-driven lighting effect. The low-resolution media façade is composed of a grid-like array of RGB LEDs that act in a fashion similar to the RGB pixels on a computer screen, by combining together to create images. By running a video signal through an image processor that maps an image to the LED array, it is possible to create a low-resolution display wall (or floor, or ceiling) with these diminutive LED lighting fixtures. Figure A-5 is an example of this kind of LED array in an interior lobby. But though color mixing with LEDs can provide a lively and entertaining lighting effect, this technique has little promise of creating energy-efficient, quality white light. How then do we use LED technology to light our high performing interior environments?

## HIGH BRIGHTNESS WHITE LEDS

The prospects for developing effective white light LED luminaires for architectural applications improved considerably in the mid 1990's with the advent of the high brightness, or high lumen, white LED. High brightness white LEDs rely on the combined luminous contribution of a colored LED, usually blue, which excites a phosphorescent material, usually yellow, causing it to glow. The process resembles the way white phosphorescent material in a fluorescent lamp is excited, and glows, to produce light. By varying the semi-conductor material and design of the LED, and the type and quantity of material in the phosphor, a white light of different color temperatures can be produced. As we've already mentioned, the efficacy of warm white LED luminaires will tend to be somewhat less than that of the cool white LEDs, partly due to the greater amount of yellow phosphor required to produce the warmer light. Figure 11-1 and Figure A-6 (in Appendix A) show examples of a high brightness white LED. There are other systems for producing white light, including the use of red or orange LEDs in conjunction with a phosphorescent coating, or the use of a phosphor-coated blue LED with some red LEDs added to the array to offset any greenish by-products. But the predominant method is to combine a blue LED with a yellow phosphor to create white light.

## LED LIGHT ENGINES

Few LED lighting fixtures rely on the output of one LED as their source. Therefore, after the LEDs are manufactured, they are often incorporated into an array made up of a group of high

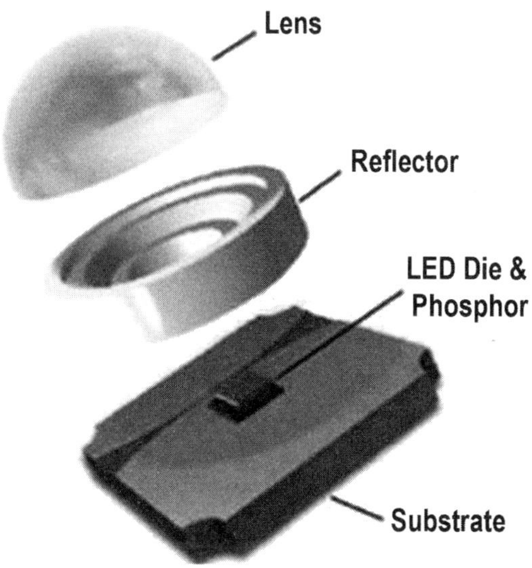


**Figure 11-1. A high brightness LED composed of a semiconductor die and phosphor, mounted on a substrate, and surrounded by a reflector cup and lens that shape the distribution of light.**

brightness LEDs all mounted together on a single printed circuit board. This array, or light engine, usually designed for a specific luminaire or luminaire type, becomes the heart of the LED luminaire in the same way a lamp is at the heart of a conventional fixture. But as we will see, there are a number of reasons why we have to think of this light engine in a different way than we think about conventional lamps. Since individual LEDs in the LED array will perform differently from luminaire to luminaire, they must be considered as of a piece with the luminaire they are integrated into, unlike conventional lamps, which can be interchanged from one luminaire to another with no effect on their performance or output.

**Figure 11-2. An array of high brightness LEDs with integral optics (lenses) configured to create a light engine for a specific luminaire design.**

## COLOR CONSISTENCY

One of the biggest challenges with LED lighting is generating a consistent color or color temperature from LED to LED, and from luminaire to luminaire. Obviously, we do not want to see an apparent variation in the color of light from the different lighting fixtures in a given space, and further, we do not want to find that, when we've replaced one of those luminaires for whatever reason, its output and color appears noticeably different than the others. This challenge is multifold. Many different steps in the production of LED chips are difficult to perform, and subject to inconsistency. To begin with, it is impossible to manufacture LED chips without variances in color. Chipmakers must resort to sorting, or binning the LEDs into different color categories. The smaller the range of colors a luminaire manufacturer wants their LED chips to fall within, the more chips must be sorted out of the bin the chipmaker can sell them, which creates more waste and results in higher costs. And there are a number of other factors that come to play in the LED, array, and luminaire design and manufacturing processes that can lead to an inconsistent color of light, or one that changes over time. A poorly applied coating of phosphor that results in an inconsistent thickness on the semiconductor die, an unstable lens material that clouds up or changes color in certain environments, a poorly designed thermal management system that results in temperature shifts across the LED circuit board, and an inconsistent drive current can all result in a luminaire whose light output exhibits a noticeable shift in color. Color consistency remains a driving concern for LED lighting manufacturers, and much research and development still goes into meeting this challenge. Many interesting strategies have been developed as a result, with at least one manufacturer making use of different colored LEDs that are automatically activated by a control circuit to maintain the overall color consistency of the light over time.

## THERMAL MANAGEMENT

A common misperception about LED lighting sources is that they generate no heat. It's true that LED luminaires use less energy to begin with, generally do a better job converting the energy they do use into light, and in doing so create a good deal less heat as a byproduct than an incandescent, halogen, or metal halide lamp. In addition, recessed LED luminaires do not tend to radiate heat into the occupied spaces where they are used. This may have the added benefit that they will create less of a cooling load in conditioned spaces than the hotter-burning sources. But they do generate heat, and that heat is detrimental to the LED chip itself, reducing its life expectancy and performance by potentially great degrees. In fact, the biggest technical challenge in designing LED luminaires is the requirement that the heat generated by the LED light engine be transmitted away from the chips as quickly and efficiently as possible. And for recessed luminaires, since the heat will generally be conducted through the back of the fixture into the ceiling plenum, where it can add to the ambient temperature there and create a potential for additional thermal stress on the luminaire, we have to consider the different conditions that may amplify this problem. For example, an insulated ceiling or a ceiling in which a radiant heating system is installed for the floor above will both add additional heat to the plenum, which would need to be considered before specifying recessed LED luminaires.

## LED PERFORMANCE AND LIFETIME (L70)

The luminous efficacy, or performance, of an LED source and its life expectancy are the product of a number of factors, most importantly the drive current and the maintained temperature of the LED chip. LEDs do not burn out like incandescent or halogen lamps. Instead they slowly degrade and their luminosity decreases over time. The lifetime of an LED source is generally measured in terms of the number of hours the LED can be used before it reaches 70% of its original light output, a metric that is also referred to as L70. Up to a point, the higher the current used to drive the LED, the brighter it will glow, and the more heat it will generate. But when the LED chip overheats a number of things will happen. There can be an appreciable shift in the color of its light, its light output will decrease, and the time it takes to reach L70 will be shortened. If the chip overheats too much, it can outright fail. So once again, the various components of an LED luminaire—individual LEDs incorporated into a light engine, a driver to regulate the drive current, and a thermal management system to conduct the heat generated by the LEDs away from the luminaire—must all be considered as parts of a larger system designed for a specific application. If the luminaire has been designed for a recessed, insulated ceiling, with a narrow aperture, it will have different thermal management requirements than if it is designed for a drop ceiling with a plenum or if it is an open track fixture. Even changes to the overall optical system, and the incorporation of lenses, baffles and reflectors to shape the projection of light will have an impact on the requirements of the thermal management system for a given luminaire. So it is impossible, at least with today's LED technology, to consider the output or lifetime of the light engine, or the LED sources themselves, separately from the luminaire as a whole. For the time being at least, fixture manufacturers have to design their LED light engines around the fixtures they create. The result is that today's LED light engines are, for the most part, of proprietary designs, with

**Figure 11-3. A recessed LED downlight. Note the heat sinks (fins) incorporated into the light engine. These are a primary component of the fixture's thermal management system. Photo credit: Cree, Inc. (c) 2011.**

no standard formats widely adopted by the industry. But there are some chipmakers and luminaire manufacturers who are beginning to develop light engines that can be incorporated by third party luminaire manufacturers into certain classes of fixture types. And it is likely only a matter of time before we see the industry develop standardized design parameters for the incorporation of a few light engine types into many different luminaires. When this happens, it may make the maintenance and design issues surrounding LED lighting a bit simpler.

## ABSOLUTE PHOTOMETRY & THE LUMINOUS EFFICACY OF LED LUMINAIRES

At the time of this writing, high brightness LEDs have been demonstrated to produce approximately 200 lumens per watt of electricity in laboratory settings. But it's a long way from the laboratory to the luminaire, and the LED arrays we are currently able to incorporate in our architectural lighting fixtures are still performing at significantly lower efficacies, and at efficacies that vary from luminaire to luminaire and application to application. By now it will be clear, for all the reasons we've just discussed, that we generally cannot quantify the performance of an LED, or an LED light engine, separately from the luminaire in which it is incorporated. If our conventional concept of a standard, modular source—a metal halide lamp, for example, that can be interchanged between different metal halide fixtures—does not apply when we're working with LEDs, then our method for calculating the light output (or the photometry) of an LED source and luminaire, and the relationship between the two, has to change as well. Instead of testing the photometry of the LED source separately, and then incorporating that data into the photometry for a luminaire, that takes into account that luminaire's efficiency, to get a result that quantifies the actual luminous output and distribution of light (a process we call relative photometry), LED lighting requires an absolute photometric method that tests the light output of the luminaire and source together. As we've noted, this is because the performance of the source incorporated within will be affected by the design of the luminaire itself, and also by the application that luminaire is designed for. This sea change in the way we calculate photometry is something we need to keep in mind when attempting to compare the luminous efficacy of conventional sources with that of LED luminaires. It is usually not possible to make a direct comparison in this way. The chart from Chapter 9 of this book (Figure 9-1) shows that the efficacy of LED downlight fixtures may still be less than that of compact fluorescent sources. Why then would we even consider LED luminaires? They are more expensive, and if we are not saving money on energy costs then how can we justify switching over to this technology? The answer comes with an understanding of the difference between absolute and relative photometry. The efficacy quoted for the LED fixture is a final calculation of how much light we'll get per watt of electricity with that luminaire, whereas the efficacy of conventional sources, or lamps, will be effectively reduced by the efficiency of the luminaire we put that source into. (See Figure 10-10 in the previous chapter.) In the final analysis, we can demonstrate an energy savings today using LED downlights in place of compact fluorescent luminaires, even though this may not be apparent when making a direct comparison between the efficacy of the LED luminaire and the compact fluorescent source. However, we *can* successfully compare the efficacy of different LED fixtures in a direct one-to-one relationship, and looking at the efficacy of one LED fixture as it compares to another in a similar application can be a valuable exercise. It should also said that the engineering challenges facing LED luminaire manufacturers are being addressed with increasing success, and the experimental, 200 lumens per watt, high brightness LED mentioned at the start of this paragraph can be seen as an embodiment of the promise of the higher and higher efficacies we can expect to achieve with LED lighting in the near future.

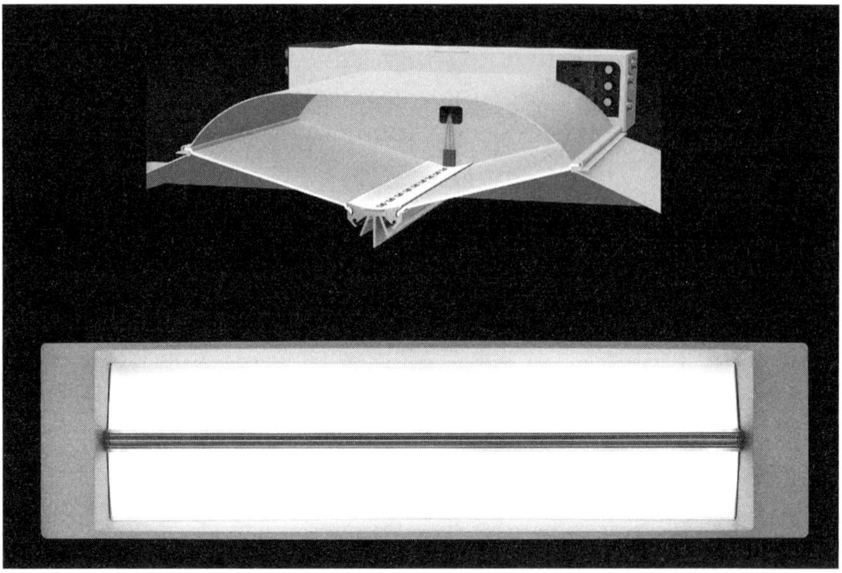

**Figure 11-4. A late design, LED 1x4 troffer fixture, with an efficacy of approximately 90 lumens/watt, designed to replace fluorescent troffers for offices and commercial use. Shown is a view of the fixture as it appears in the ceiling (bottom) and a cutaway of the interior of the fixture (top) that reveals a system of linear LEDs in an indirect application to create a large, luminous source. Photo/Image credit: Cree, Inc. (c) 2011.**

## NEW PHOTOMETRY, NEW STANDARDS

With the shift from relative photometry to absolute photometry that SSL lighting requires, the IESNA has developed new standards to govern the testing and rating of both LED chips and LED luminaires. Prior to the release and eventual adoption of these standards, many luminaire manufacturers were simply reiterating the LED chip manufacturer's performance data for the chips incorporated into their luminaires as indicative of the performance of the luminaires themselves. This was a misleading and in some cases dishonest practice that falsely raised the expectations of consumers purchasing these early LED luminaires. And it did a lot of potential damage to the credibility of the LED lighting industry. Even today we occasionally see unscrupulous luminaire manufacturers misquoting an LED chip manufacturer's performance data in this way. Now, with the development and widespread adoption of new absolute photometric testing standards, this behavior is, thankfully, seen far less frequently. Today, all the reputable companies manufacturing high quality LED luminaires publish performance data based on the testing of their luminaires to the specifications of these new standards. In addition, they are using LED chips from chipmakers that test their chips and arrays similarly. ***IESNA LM-79-08*** is the approved method for testing **complete luminaires** that include LED chips, electrical components (drivers), and thermal management systems (heat sinks). The results of these tests include: total light output, lumen distribution, input power, luminous efficacy, color temperature, and CRI. Experimental controls are specified in this standard that relate to ambient conditions like temperature, mounting, and airflow, as well as the characteristics of the power supplies used to drive the luminaire, and the instrumentation employed in the testing. ***IESNA LM-80-08*** is the approved method for testing the lumen depreciation of **SSL sources** (the LEDs themselves, individually or when combined into modules and arrays) outside the context of their integration into a luminaire. The testing takes place over a prescribed period of time, at controlled temperatures, with measurements taken at regular intervals. Results include: lumen depreciation, color shifts over time, and failure rate. Currently, luminaire manufacturers extrapolate the life of their LED fixtures based on the results of LM-79-08 and LM-80-08. Both of these standards specify testing methods, and are not intended as a seal of approval. But when reading the photometric reports your luminaire manufacturer provides, it is good to be able to verify the means employed to generate them. **One good place to start when evaluating an LED luminaire is to ask the manufacturer if the photometric specifications were developed via the testing procedures specified in LM-79, and if the chips employed in the lumi-**

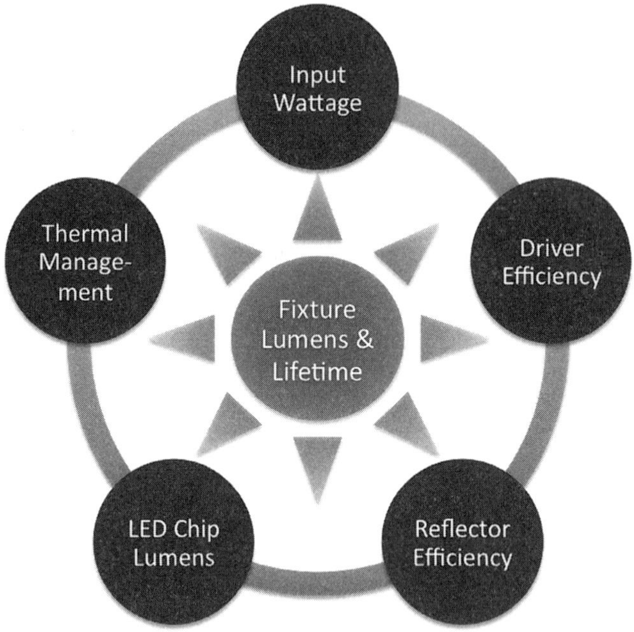


**Figure 11-5. Absolute photometry: It's a system! The drive current, thermal management and fixture design all combine to define the performance and life expectancy of the Solid State luminaire in a system in which the source and luminaire can not be considered separately.**

**naire were tested via the procedures specified in LM-80.** Many of the better luminaire manufacturers will state this clearly on their specification sheets.

## LCA & MODULARITY

LED sources have a very long life expectancy. Most of the well-made luminaires come with a projected lifespan of 50,000-60,000 hours before they reach L70 (longer in colder environments). This makes LED a naturally good choice from the point of view of a life cycle assessment. Obviously, the longer a product lasts, the fewer have to be manufactured, transported, and sold, and the fewer need to be replaced and (hopefully) recycled so that the materials contained therein can be recovered and reused. Or, in the worse case scenario, the longer a product lasts, the fewer will be dumped into a landfill! But if the LED luminaire is of a piece with it's source, i.e. if LED sources are so different from conventional lighting sources that one can't simply replace them, as one would a light bulb, when they've reached the end of their useful life, then does that mean we have to discard the entire luminaire when the LEDs within have reached L70? And if we have to discard the entire luminaire when the LEDs have reached the end of their life, even if its life is 50,000 hours, is that a sustainable solution? The answer to the last question is, no. Unfortunately, not all SSL luminaires are manufactured with a suitable accommodation for this piece of the sustainability question. To be sure, there are some significant engineering challenges inherent in devising systems whereby the LED chips or arrays are easily separable from their thermal management systems in the field, by the end user. And it's due to these challenges that the first generation of LED luminaires, and even some of today's less finely-engineered fixtures do need to be entirely replaced when the LED sources within reach the end of their life. Subsequent generations of fixture designs have seen the development of luminaires that separated the "light engine" from the rest of the housing, allowing for the replacement of the LED array, electronic driver, and heat sinks. But this is still a lot of material to discard. The current drive in the market is to develop replaceable light engines that constitute smaller and smaller amounts of material, so as to generate less waste over the long run and to maximize the useful life of the other components of the luminaire. The better SSL luminaire manufacturers are addressing this problem with a good deal of success, and in some cases have brought to market fixtures where the LED arrays—the individual LEDs on their printed circuit boards—are replaceable as a totally separate element from the fixture housing, electrical driver, LED optics, and thermal management system. In doing so they are also future-proofing these luminaires to be able to

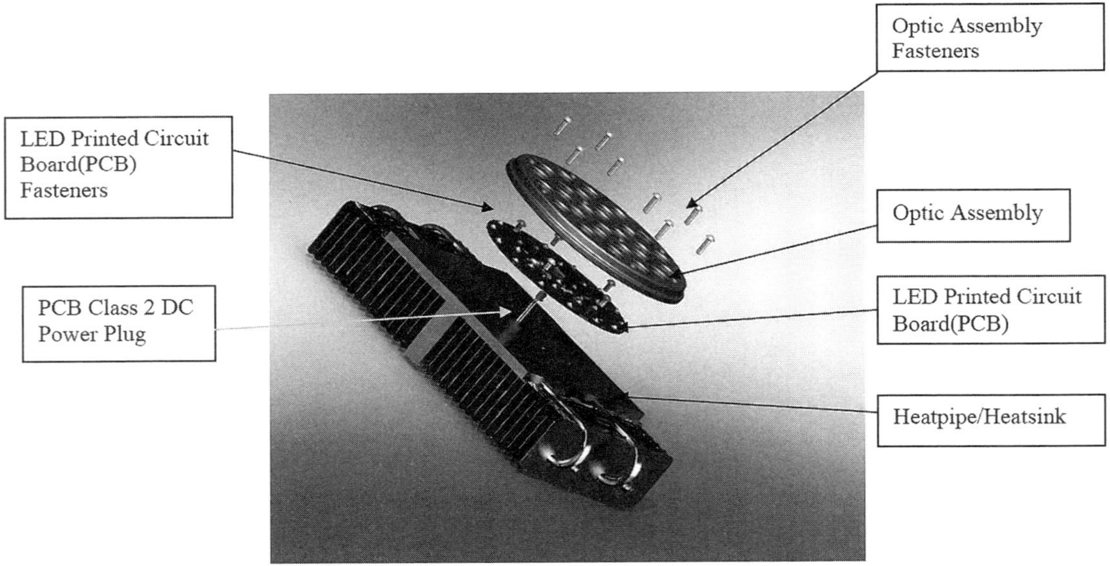


**Figure 11-6. A replaceable LED light engine (printed circuit board) that is separable from the luminaire's electrical components and heat sink and provides for in-the-field substitution of optics, enabling the user to change the light distribution that is projected from the fixture or substitute the engine for a future model that might utilize fewer LEDs to deliver the same amount of light. Photo courtesy of Ruud Lighting Inc. (BetaLED®).**

accommodate higher performance arrays that may deliver the same amount of light with fewer LEDs when the original light engine is replaced. We are also seeing a number of manufacturers developing standardized light engines that can be incorporated into different luminaires, designed and manufactured by different companies. These modular SSL light engine manufacturers are doing the most to reinvent the way we conceive of a light bulb, and at the same time creating a set of standardized "lamps" for the new and future generations of LED lighting fixtures.

## LED REPLACEMENT LAMPS

Though there are an ever-increasing number of LED retrofit lamps on the market, by now it may be apparent that quality LED lighting does not generally come in the form of lamps that are designed to fit into fixtures that are designed for conventional sources. On some level this is unfortunate. Everyone would gain from a new technology that allowed us to make better, more efficient use of the existing lighting infrastructure. The ability to switch to SSL sources by simply changing light bulbs would be beneficial indeed! As LED chip designs improve, the likelihood is that lamp manufacturers will develop cost effective LED lamp replacements that work well in conventional lighting fixtures. (Some of the largest lamp manufacturers, themselves divisions of very large corporations, are right now hard at work on this very challenge.) But for the time being, with a couple of notable exceptions that we will discuss later in this section, this is not the best way to apply the currently available LED lighting technology.

In the same way that we must consider an LED luminaire and its source as a complete system designed for a specific application, we need to consider an LED replacement lamp as an LED luminaire unto itself, and not as a conventional lamp. LED replacement lamps are complete systems, and should be tested according to the IES LM-79 standard, as described above. They incorporate LED chips, reflectors, lenses, and drivers with a thermal management system, and are designed for specific applications. Some are designed to work in enclosed, recessed, fixtures, and some will only work well in open fixtures that allow for more air circulation around the lamp. They cannot be inserted into any existing fixture with the expectation that they will always perform in the same way, as different fixture types will create different conditions that may restrict that particular "lamp's" thermal management, potentially causing it to perform less well or even shorten its life. Most of the good-quality LED lamps on the market will come with a written specification that will indicate whether or not the lamp can be used in an enclosed fixture and the maximum ambient temperature in which the lamp can be expected to operate properly, as well as complete photometric data developed via the testing standards specified in IES LM -79. In the absence of a published specification, the manufacturer should be able to provide this data when queried. If they cannot, their products should be avoided.

In addition to having a basic understanding of the proper way to evaluate the performance of these devices, there are a number of questions that one should consider before specifying LED replacement lamps for existing luminaires. The first is: what are the photometric properties of the LED lamp and are they appropriate to the luminaire it is being used in? In the case of an LED replacement for a T8 fluorescent lamp, for example, one should ask what will be the result from putting a directional source into a fixture that is designed to collect and distribute the light from an omnidirectional fluorescent lamp, which emits light in all directions, and is the output of this replacement lamp an equivalent to the original source. The actual form factor of the lamp must also be considered. Will the so-called LED MR16 or PAR 30 replacement lamp, with its surround of heat sinks, actually fit into my existing MR16 or PAR 30 fixture? Finally, when considering an LED replacement lamp one has to ask if the manufacturer is reputable and if they are publishing reliable or misleading specifications regarding their products. The unfortunate fact is that many of the LED replacement products that have saturated the market are of dubious quality, with low output and poor color rendering. One good resource that gives a broad view into the general state of the market in various categories of SSL replacement lamps is the US department of energy's *Commercially Available LED Product Evaluation and Reporting (CALiPER)* program. This program has been tracking the development of SSL lighting over the past few years, and is a valuable tool for evaluating the quality of the many commercially available SSL products, including replacement lamps. Reports and summary evaluations from this program can be found at the following website:

http://www1.eere.energy.gov/buildings/ssl/about_caliper.html

**LED Reflector Lamps (a notable exception)**

LEDs, by their very nature, lend themselves to the design of directional sources. For this reason the most successful LED replacement lamps are those patterned after our conventional re-

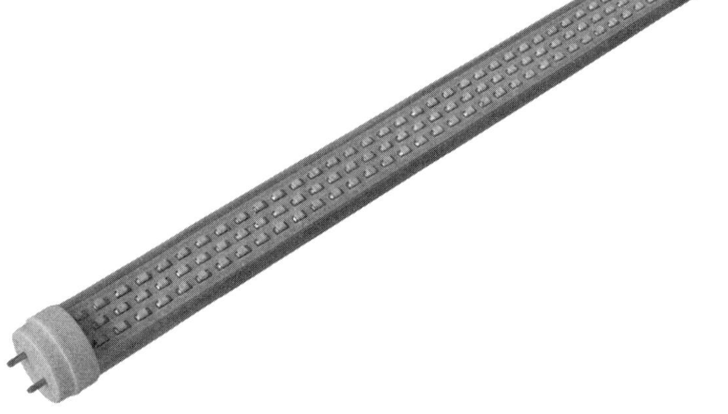

**Figure 11-7. Will the optical distribution of this directional LED replacement "lamp" give a good result when incorporated into a fixture designed to collect and distribute the light from an omnidirectional fluorescent lamp?**

flector lamps. For the time being anyway, the larger LED "reflector lamp" formats are better than the smaller ones, with some very good PAR 30 and PAR 38 replacement products available, though the good ones are by no means inexpensive. In this arena one must follow the age-old advice: "If it seems too good to be true, it is." At the time of this writing, there are no high-performing MR16 LED replacement lamps on the market, but it is likely just a matter of time before there are. And it is also important to note that some low-voltage transformers that currently power MR16 systems, and the dimmers that control them, may not work with an LED replacement, as the wattage of the LED lamp is lower than what these devices are designed for. Just as LEDs are suited to directional applications, like reflector lamps, they are not a natural replacement for sources with an omnidirectional distribution of light, like an incandescent or fluorescent lamp. And so they do not tend to work well when configured into legacy lamp formats that fit into existing fixtures designed to maximize the luminous distribution of these conventional lamp types. With so many millions of existing fixtures designed around the incandescent "A" lamp and linear fluorescent tube currently installed around the world, it is understandable that lamp manufacturers would want to exploit such a large market for more efficient lamps in these formats. But designing LED

**Figure 11-8. A good quality LED PAR lamp. These lamps should be considered as complete luminaires: designed for a specific application and incorporating thermal management systems, electronic drivers, reflectors, and lenses along with the LED source. Many of these are only designed for unenclosed applications, and are not for use in recessed fixtures. Photo credit: Cree, Inc. (c) 2011.**

lamps to fit into legacy formats is somewhat counter-productive, and may not provide the most efficient result, as it is an attempt to wrap a new technology, with very different properties than the technology it is replacing, in an envelope specifically designed for that old technology. That said, with so many legacy lighting fixtures in use today, the incentive to do just this is great. And as LED sources become more efficacious, the day will certainly come when lamp manufacturers can create inexpensive, good-quality, LED lamps that fit into all of our old lighting fixtures. But for the time being anyway, this will remain a "holy grail" for the LED lamp industry to pursue.

## LUMEN DISTRIBUTION & LEDS (A NATURAL FIT FOR A SUSTAINABLE DESIGN STRATEGY)

In addition to the ever-increasing efficacies of LEDs, another key factor contributing to the potential for their success as energy efficient lighting sources is their suitability, due to their very small size, for the creation of highly controllable luminaires with exacting photometric distributions. This essentially means they are very good at getting the light where we want it, and in doing so wasting very little of the luminous energy they emit. As we've discussed in the section on relative photometry, most conventional sources create an omnidirectional distribution of light. That is to say, a fluorescent tube and a conventional light bulb emit light in nearly all directions. However, a good many of the applications where we require electric light do not lend themselves to an omnidirectional distribution. The direct/indirect linear fluorescent pendant is one notable exception, and there are others. But a significant number of the luminaires we employ rely on some kind of directional re-distribution of the source's light to satisfy the requirements of the application they were designed for. Downlights, wall washers, wall grazers, roadway lighting, sign lighting, cove lighting, uplighting, site lighting, and many others, all require the projection or distribution of light to be shaped in one way or another. And unlike fluorescent, incandescent and most larger, higher-lumen tungsten and metal halide lamps, LEDs lend themselves, out of the box, to the design of directional luminaires. They produce a natural distribution of light that is often about 120 degrees (versus a nearly 360 degree distribution with most conventional sources) and they are easy to integrate with optics (lenses) to create a variety of narrower projections with a minimum of loss, giving us a higher number of *effective lumens* per watt of electricity used. We can now manufacture minute optics with projections that may be symmetrical (like a conventional accent light) or asymmetrical (like a wall washer) without the use of a reflector that in redirecting the "back end" of a lamp's output will often generate significant inefficiencies. These factors all contribute to the promise of LEDs as a truly efficacious source for many of our everyday lighting applications.

# Part IV

# Sustainable Applications: Daylighting & Lighting Controls

# Chapter 12

# Daylighting

## WHAT IS DAYLIGHTING?

Daylighting is the practice of placing windows or other openings in relation to reflective interior surfaces in a building so that, during the day, natural light provides *effective* interior illumination. Particular attention should be given to incorporating daylighting techniques when designing a building where the aim is to create a healthy, productive, visually comfortable, and inviting environment while simultaneously reducing energy use. There are many clear advantages to incorporating daylight into interior spaces. One of the most obvious is that the more we use natural daylight, the less we will need to rely on electric lighting, and the more we can reduce energy use and minimize our carbon footprint. But daylighting brings advantages beyond satisfying our need to live and work in a more environmentally sustainable way. The psychological and physiological benefits of daylighting have been the subject of many studies, including some that examine the positive activation of the human circadian system by daylight, and others that concentrate on a presumed innate tendency to want a connection with the outside (natural) world. Though the scientific jury may still be out as to the specific reasons why, it is nonetheless widely accepted that having access to natural light is desirable, and that it can contribute to, and enhance, human health and wellbeing. [12]

It is self-evident that people want access to daylight, and it's no accident that in the workplace this access has traditionally been allocated to those who inhabit the upper levels of the socioeconomic strata. In our commercial and corporate cultures the corner office, with windows on two sides, is identified with upper management, and the interior, windowless spaces are often assigned to clerical and back office workers. Why? Because access to daylight is highly valued, and has therefore been traditionally reserved for executives, managers, and others who occupy higher positions in the workplace. There is evidence that regular exposure to daylight contributes to keeping us healthy, happy and productive. Yet many of us spend almost 90% of our lives indoors. [13] This is an unavoidable fact of life in today's world, but through the proper implementation of daylighting we, as designers of high performance buildings, can help ameliorate this unfortunate modern-day condition, and in doing so we can add the savings of a significant amount of energy (and money) into the bargain.

## EFFECTIVE ILLUMINATION

Effective illumination is an important concept to understand in the context of designing successful daylighting treatments. Not all illumination is effective, and more light is not always better light. In fact, too much light in the wrong place can be blinding. Illumination is effective when it facilitates increased productivity and supports our ability to accomplish the visual tasks required in our everyday lives. But when a great quantity of light infiltrates a less-brightly lit environment, it can create high contrast ratios between surfaces, or high luminance ratios, which can in turn cause glare. And daylight, especially direct sunlight, is made of a great quantity of light indeed! In Chapter 7, where we discussed target illuminance levels, we identified the range between 10 and 50 footcandles as being a sufficient amount of illuminance for interior living and working spaces for the majority of people engaged in most daily activities. And so it follows that most of the interior lighting in our homes, offices, and stores will provide footcandle levels that are generally within this range. But full daylight gives us illuminance levels as high as 1,000 – 2,000 footcandles, and direct sun can deliver illuminance levels in excess of 10,000 footcandles. That's at least 200 times the amount of light that is present in most indoor, electrically lit environments. As previously noted, luminance and illuminance are directly related to one another. The luminance of a surface is, essentially, the amount of light falling on or passing through that surface—the illuminance—that is reflected back to or directed towards the viewer. The higher the illuminance that is present, the higher the luminance of a surface, and the brighter that surface appears. We can see perfectly well where the illuminance may be 5 footcandles or 5,000, and we can view many kinds of surfaces with a wide range of luminances without experiencing glare, but our eyes do not have the capacity to accommodate so great a range, or such a high degree of contrast, all at the same time.

Clearly then, when direct sunlight enters a building that is otherwise lit with electric lighting and falls on some surfaces, the result can be a high-contrast, glary condition that can have the potential to hinder our ability to function, and make us downright uncomfortable. In addition to this, the windows themselves can have a high luminance and create glare. This is all complicated by the fact that what constitutes a glare condition can be quite different from individual to individual, and, interestingly enough, studies have shown that people will tolerate a higher degree of contrast between surfaces while indoors when it is due to the inclusion of a window in their field of view. Furthermore, the quality of the view seen through the window can raise their tolerance for even higher contrast, or luminance, ratios. [14] This seems to indicate that people are willing to put up with some discomfort so that they may have access to daylight. Still, the sheer intensity of daylight can be such that overly high contrast ratios and glare conditions will often result unless we take steps to mitigate them. For daylighting to be effective then, we must take pains to provide a degree of control over the luminance of the view windows, and we must minimize the infiltration of direct sunlight onto any work surfaces, or any surfaces in our general field of view. We can do this by incorporating architectural features and materials that redirect or diffuse the light so as to create a soft, indirect distribution that better penetrates the interior of the space. This is a concept we've recognized for some time, evidenced by the regular the adoption of blinds, shades and curtains long, long ago. In a way, the simplest forms of effective daylighting have been practiced for centuries. As soon as people cut holes in their dwellings to create windows, and then incorporated shades and curtains to block or diffuse the light when it was overpoweringly bright, they began practicing rudimentary forms of daylighting.

Creating effective illumination with daylight, like lighting design itself, has an element of art to it, but there are some definable strategies we can employ. An integrative design process for new construction that considers daylighting will start with the building's siting and orientation and include the placement and sizing of windows, skylights and other daylight apertures,

**Figure 12-1. An early method of daylighting: The window reflector was used to provide indoor illumination to supplement poor quality or insufficient interior lighting in offices, and other commercial spaces. Image credit: Otto Lueger, Lexikon der gesamten Technik (Dictionary of Technology), 1904.**

as well as the color and reflectivity of the interior surfaces. In many cases, northern exposure will be desirable, so as to maximize the indirect daylight allowed to enter the space, while minimizing direct sunlight infiltration. Other elements to be considered include architectural treatments designed to reflect and push the daylight further into the interior space, shades, louvers and other devices to control and reduce the amount of direct sunlight that can enter as the exterior conditions change, and the division of the space itself. In many cases we are seeing the traditional office plan, with private offices for managers and executives at the perimeter, that block access to daylight for the support staff, replaced by fully open offices, or glazing between the perimeter offices and the workstations at the interior, to allow the daylight, and views, to reach a greater proportion of the workers. Successful daylighting strategies often make the best use of natural light, especially direct sunlight, by redirecting it towards reflective overhead surfaces, creating illuminated planes that project the light farther into the building's interior. It's important to keep in mind that, like electric lighting systems, daylighting systems that are not effective will often be retrofitted in ways that are counter to the original intent of the designer. South facing widows without operable solar shades or blinds, that allow direct sunlight to enter, may end up being covered entirely by the occupant. Successful daylighting is not about creating as many widows as possible, or utilizing skylights in a willy-nilly fashion. Care must be taken to regulate the daylight infiltration, as too much light in the wrong place, or from the wrong direction, can be as detrimental to the quality of an interior environment as too little.

## ARCHITECTURAL DAYLIGHTING TREATMENTS

Architectural daylighting treatments come in many forms, from discrete additions to a building's structure that are specifically designed to carry daylight into the interior spaces, to design choices integral to the very shape of the building itself, like the one shown in Figure 12-2. The most effective of these will work by directing daylight to matte, reflective surfaces to create an ambient light that is as soft, shadow-free, and omnidirectional as possible.

Well-placed skylights and windows in conjunction with diffusing shades and blinds can go a long way toward providing effective natural illumination. Additional architectural elements can also be incorporated to maximize the daylight penetration. Some of these are quite simple, and sometimes available as a retrofit to existing structures. One of the most common is the light shelf, a shelf directly below a clerestory window (a window well above eye level, usually right under the roof line) extend-

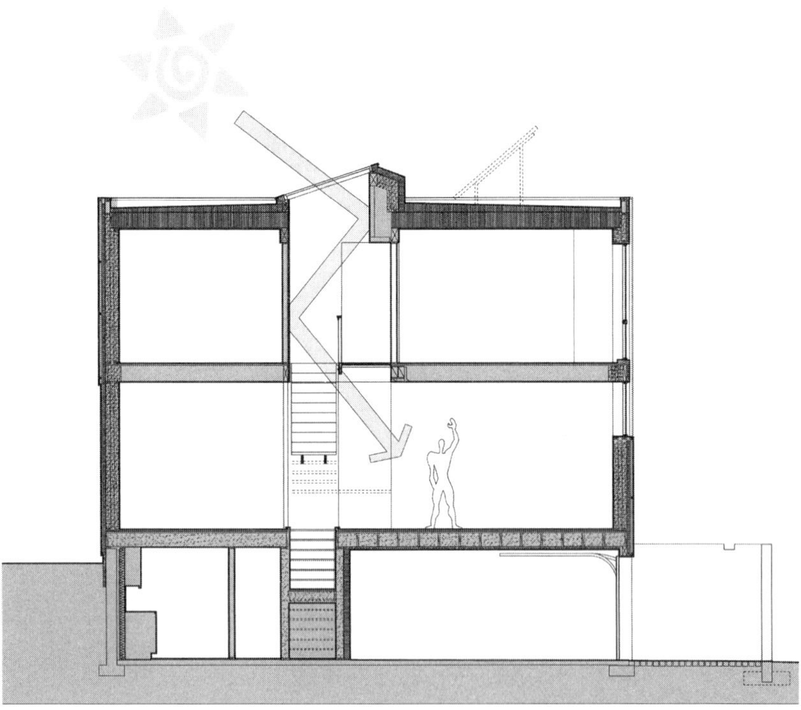

**Figure 12-2. A section showing a residence with a central daylighting treatment integrated into the building design. In this case a glass floor is planned for the second floor hallway, to allow as much daylight to pass through to the first floor as is possible. Image credit and design: Heliotrope Architects.**

ing to the exterior and interior of the building. These are usually finished in a light color so as to maximize their ability to collect and reflect daylight, be it direct sunlight or indirect light from the sky. A light shelf will reflect light up and into a light colored interior ceiling where it can then be redirected as soft, indirect light, deeper into the room than would otherwise be possible. Light shelves will usually separate a clerestory window from a lower view window, and can also provide relief from direct sunlight that might otherwise infiltrate the space through the view window (see Figure 12-3).

It is always best to model architectural daylighting treatments as completely as possible, as a variety of seemingly innocuous details affecting the shape of the space can have a great

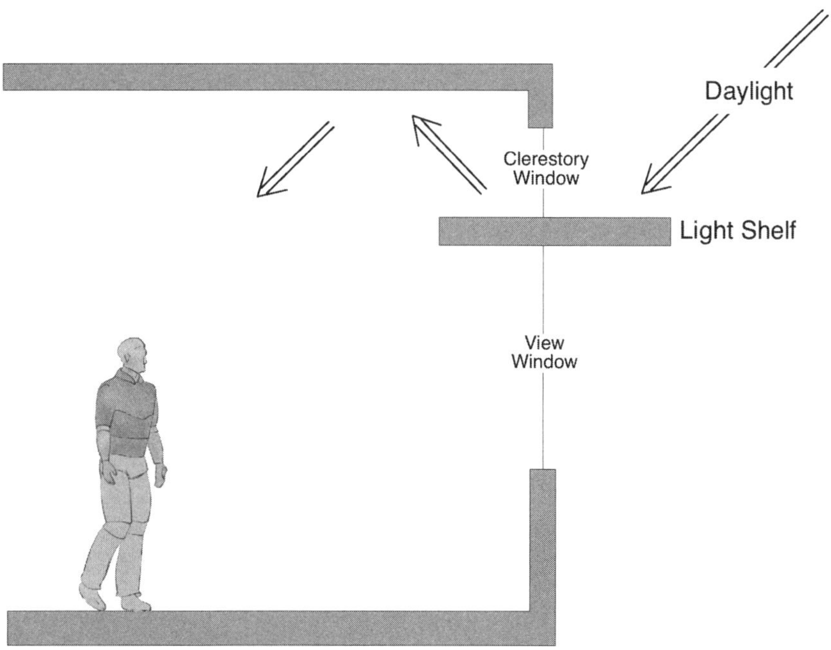


**Figure 12-3. A light shelf is a simple way to bring effective daylight into a space by reflecting it into a light colored ceiling where it can be redirected as a soft indirect light to the interior space; while simultaneously blocking direct sunlight from entering the space through the view window.**

effect on the efficacy of the proposed treatments. For example, light shelves generally work better in higher spaces and in conjunction with clerestory windows that are themselves of a greater height. And in some cases, a small discrepancy along these lines can make a big difference. Figure 12-4 shows renderings and calculations for two very similar rooms at the same location and the same time of day with one of the rooms having a slightly higher ceiling, enabling the designer to accommodate a clerestory window height that is commensurately greater. The increase in the illuminance achieved in the space from this seemingly small difference is considerable. Furthermore, the study showed that the benefit of incorporating a light shelf in the lower ceilinged room was inconsequential versus having no light shelf at all, leading

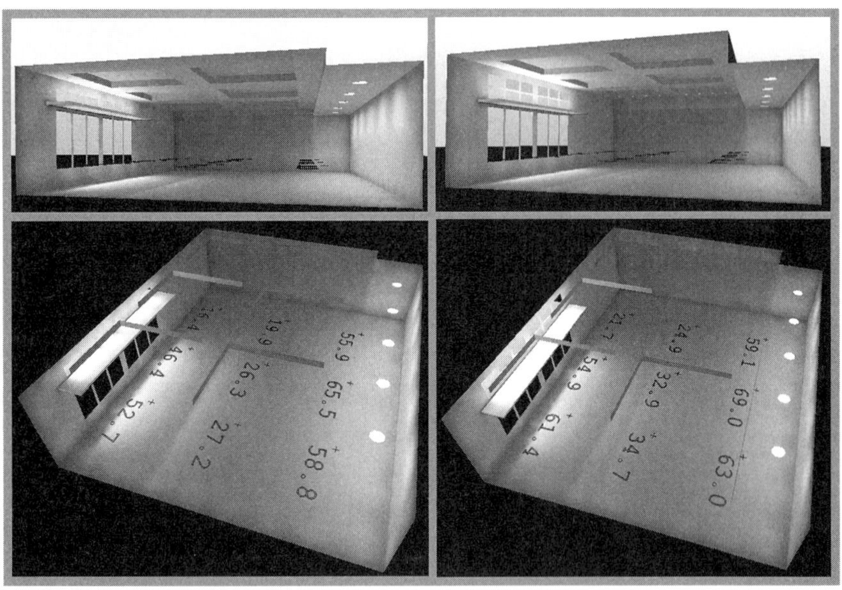


**Figure 12-4. Light shelves are most effective when employed in higher ceilinged spaces and with as high a clerestory window as possible, as shown in this daylight study of two nearly identical rooms at the same time of day and in the same location. The room on the left has 9′ ceilings and a 1′ clerestory window, and the room on the right has 10′ ceilings and a 2′ clerestory window. Projected illuminance levels are shown, in footcandles, and are significantly higher in the case of the slightly higher-ceilinged space.**

to the conclusion that the minimum ceiling height in this space for a light shelf to be even worth consideration would be 10′, with a 2′ high clerestory window.

Another technique that can be used to great effect is the sawtooth skylight, usually oriented to the north to collect indirect light from the sky. When incorporated into a properly sited building, this design will exclude direct sunlight and admit indirect daylight, reflecting it off of the canted ceiling planes and into the space below. This is a very effective treatment for large spaces, such as industrial facilities, large retail stores, and transportation hubs. In cases where the building cannot be sited such that the skylights face north, the inclusion of operable solar shades to diffuse any direct sunlight entering the building is necessary. There are many other skylight designs that can be utilized to create effective daylighting systems, and many ways to manage the direct light that might otherwise infiltrate the space: from simple louvers to complex, computer modeled cellular baffles designed to block the direct sun from entering a skylight on a particular building in a particular location. Some architects make a practice out of designing their buildings to make the best use of daylight, and if daylighting is considered from the beginning of the design process, the permutations and possibilities are many.

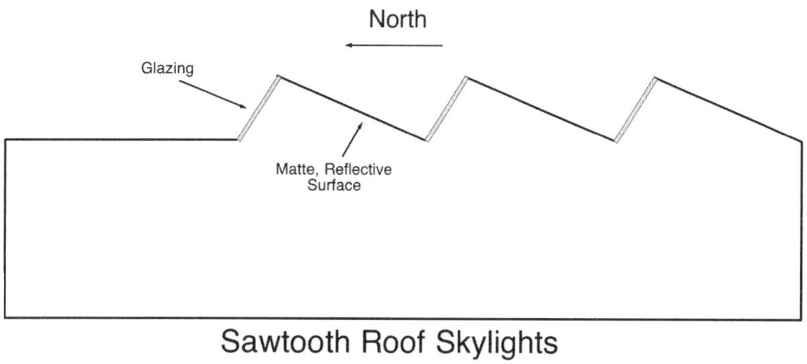


**Figure 12-5. Sawtooth roof skylights, oriented to the north, provide excellent indirect daylight to the interior space. The inclusion of operable solar shades, to diffuse direct sunlight, makes this an effective daylighting treatment in cases where the building siting is such that the skylights do not face to the north.**

# Chapter 13

# Lighting Controls

## THE CASE FOR ENHANCED LIGHTING CONTROLS

Lighting controls are a central component of any high-performing, energy-efficient lighting solution. Controls allow us to dim the lighting to our preferred level, or to the level required for a particular activity or task. Controls turn electric lighting on and off per a pre-determined schedule or based on room occupancy. With the right controls we can have the lighting in our workspaces automatically dimmed in response to a request by the utility company to decrease demand for electricity during periods of peak usage, or in response to the amount of daylight present in the room. All of these are scenarios where lighting controls help us conserve energy by automatically limiting the amount of electrical power used by our lighting systems when it is appropriate to do so.

Lighting controls of the many types we will discuss in this chapter all constitute proven energy conservation strategies. But the value of lighting controls has not always been reflected in our national and state building energy codes, with the exception of some requirements for: occupancy sensors, manual occupant controls (i.e. readily accessible light switches), automatic shut-off controls in interior spaces of a certain type and size, and scheduled shutoff of exterior lighting during daylight hours. Until recently, most of our building standards and energy codes relied on a lighting energy-efficiency strategy whose central component was reducing the connected load of the electric lighting systems. *Connected Load* refers to the amount of power the lighting system would draw if all the lighting were turned on at the maximum level possible, using the maximum wattage lamps allowed for each lighting fixture. This approach is based on the concept of regulating the maximum amount of energy required to power a lighting system, but does not take into account, or have any positive effect on, the way we actually use this system. When looking at the connected load of a lighting system we are only examining the amount of energy that *could* be used, not the amount that actually *is* used. It's analogous to defining the minimum miles a car must be able to travel on a gallon of gasoline, without considering the energy that may be wasted when we leave that car's engine idling in our driveway for hours at a time.

Codes and standards will be covered in greater detail later in this book, but for now it will be enough to explain that the building energy codes in much of the United States use an installed lighting system's connected load to determine if that system complies with the lighting power allowance for a building or space. These lighting power allowances are based on a calculation that takes the area of a certain type of building or interior space and multiplies it by the prescribed lighting power density (LPD) for that building or space type. The LPD is expressed as a certain number of watts per square foot, and it represents the maximum connected load one is allotted for the lighting system per square foot of a particular space. So if we are allowed a lighting power density of one watt per square foot for general lighting in an open office, and that office has an area of one thousand square feet, then we will have a lighting power allowance of one thousand watts for the connected load of that office's general lighting system. Since all spaces do not need to be lit in the same way, the allotted LPDs are higher or lower for different kinds of spaces, allowing the use of more power for brighter lighting in areas where more complex tasks are to be performed. Certain manufacturing facilities and healthcare clinics are examples of spaces where higher LPDs are allowed. Lobbies and corridors, where it's generally agreed lower illuminance levels will be sufficient, are examples of spaces where the code might specify lower LPDs.

On the face of it, this is a perfectly reasonable strategy to regulate our energy usage for lighting, just as it's reasonable to assign a minimum miles-per-gallon standard for various kinds of automobiles. But there is a limit to this approach. Taking the automobile analogy a bit further, if we simply tighten the minimum miles-per-gallon requirement with no regard as to our capacity to actually produce cars and trucks that run on less and less fuel, we may end up adversely affecting our ability to get around, or run our businesses. As an added measure, and in place of developing unrealistic miles-per-gallon standards, we can also limit the amount of time those automobile engines will actually run, and by extension the amount of gas they will consume, by regulating the amount of time they can be left idling when not in motion. Indeed, anti-idling regulations have cropped up in Canada, Europe, and a few states and municipalities in the U.S. to this effect, that limit the amount of time a vehicle engine can be left running when the automobile is parked. (The above analogy should in no way be taken as a position against increased miles-per-gallon standards in the United States.)

In the same way, the necessity to increase our buildings' energy efficiency with each new iteration of our national and state building energy codes (and the standards they are based on) has resulted in such a squeeze on our allowed lighting power densities as to make it increasingly difficult to design and specify quality lighting systems. As we've discussed, the requirement to minimize glare from and manage the luminance of our lighting systems, and the general desire to create visually comfortable, and even visually interesting environments, will often result in higher connected loads than if these goals were not considered. The use of indirect lighting, and the integration of diffusers and shields in luminaires, all central components of any quality lighting design, will invariably result in the use of more energy than a strictly utilitarian, direct lighting approach, where lamps are unshielded and the focus is to simply provide a specific amount of illuminance on the work-plane. And the desire to use lighting to create visual interest, by varying the luminances of different surfaces to reinforce an existing architectural hierarchy, can sometimes require the use even of more energy than would otherwise be necessary. These are quality of life issues, and they need to be taken into account. But it's also true that limiting our energy use, especially as long as it comes primarily from non-renewable, carbon-based sources, is extremely important. And this is where advanced lighting controls can play a role as a value-added strategy to maximize our efforts to conserve energy while allowing us to maintain a high degree of quality in our interior lighting. Systems that turn the lights off when a space is unoccupied, or when they are not needed, are a good first line, easy-to-implement strategy. Integrating advanced controls that regulate the extent to which lighting is used at a full, non-dimmed level when it is on can result in even greater energy savings. Together these different kinds of lighting controls can do far more to reduce our energy use than simply reducing the allowable connected load of our lighting systems.

This realization is reflected in more recent energy standards and model building codes, the consensus-based documents created by professional organizations that municipal building codes are often based on. In fact, the 2010 release of the American Society of Heating, Refrigeration and Air-Conditioning Engineers (ASHRAE) 90.1 standard, upon which most of the commercial building energy codes in the United States are based, takes the recognition of the value of advanced lighting controls to this next step. In the latest version of this very influential standard, the inclusion of advanced controls in a project will actually trigger the allowance of higher LPDs for that project's qualifying spaces. In this case, advanced controls refers to a variety of strategies that allow for a more specific tuning of the light level to suit the changing requirement for electric lighting at different times within a space, including manual dimming control, automatic daylight harvesting dimming, and multilevel occupancy sensors. This is a bit of good news for those of us who care about the quality of our lit environment. With all of the lighting control systems available to us, we can create highly productive and pleasant spaces that are also frugal in their energy-use, but it will fall to the lighting designers who volunteer on the committees that write these codes and standards to push for a greater reliance on these technologies. There is little doubt that through judicious use of lighting controls we can design systems that will result in greater energy savings while also advancing the cause of quality lighting to a greater degree than would otherwise be possible by simply relying on the adoption of ever-more stringent lighting power allowances.

## DAYLIGHT HARVESTING

Daylight harvesting, or daylight dimming, refers to control systems that automatically dim a building's interior lighting in response to increased daylight infiltration, essentially turning the lights down when they are not needed because of the presence of an adequate amount of natural light in the space. This technology can be incorporated along with integrated architectural daylighting treatments specifically designed to maximize the effective daylighting of a particular space, or simply installed where enough light enters through the windows to effectively light that immediate area. There are a couple of different rules of thumb you can follow to help decide if daylight dimming will be beneficial to your project. Generally, if the area of fenestration (windows) is equal to or greater than 5% of the area of the overall floor space, then daylight dimming should be considered. If you have skylights throughout the space, then the benefits of daylight dimming will always be worth the additional costs.

Most of the time, as we go about the day, we don't notice when the daylight in our offices and classrooms has brightened the space to the point where the electric lighting is no longer making a significant contribution. We might not even notice that the lights are on, and as a result might forget to turn them off when we leave. With daylight dimming, as the day progresses and the interior spaces are lit with daylight the electric lighting is proportionately dimmed, resulting in reduced waste and increased energy savings. Daylight dimming is available in a wide variety of systems of greater and lesser complexity that can often be scaled to the needs of a particular project. There are large-scale, multi-zone systems that integrate with whole-building energy management controls to coordinate daylight dimming with other lighting controls, shade controls, and HVAC controls. And at the other end of the scale, there are individual luminaires equipped with standalone photosensors and dimming ballasts designed to control just themselves, for easy installation in areas where there is a great degree of daylight, and no additional automated controls requirement.

Since daylight seldom penetrates all parts of our interior spaces in equal quantities, care must be taken to insure that the lighting fixtures are controlled in groups, or control zones, that correlate to the areas where there is more or less daylight adding to the interior illumination. A control zone refers to a group of lighting fixtures, or an area in which the installed lighting fixtures can be controlled—in this case dimmed—all together. As the daylight entering interior spaces will always be greater nearest the windows and skylights, it's important that daylight dimming be correctly zoned, so that the lighting in these areas, where there is more daylight can be dimmed separately from the lighting in areas where there is less, to achieve as uniform a level of light across the entire space as possible. Except when using fixtures with integral daylight sensors that are essentially self-controlled luminaires, where each will respond a little differently depending on their proximity to the daylight source, the correct zoning of a daylight dimming system is imperative. A general rule of thumb where windows are providing the daylight is that the first daylight dimming zone should control all of the lighting fixtures within fifteen feet of the windows. In most cases where the ceilings are at a standard height, and the windows reach up to seven and a half feet; this first zone will be sufficient. In spaces with higher ceilings and windows, a second zone should be created to control all lighting fixtures between 15 and 25 feet of the windows. (See Figure 13-1 for an illustration of these zoning parameters).

Where skylights are concerned, there are somewhat different criteria used to define the daylight dimming zones. A general rule of thumb is to calculate the dimensions of the first daylight dimming zone by adding the height of the ceiling to the width

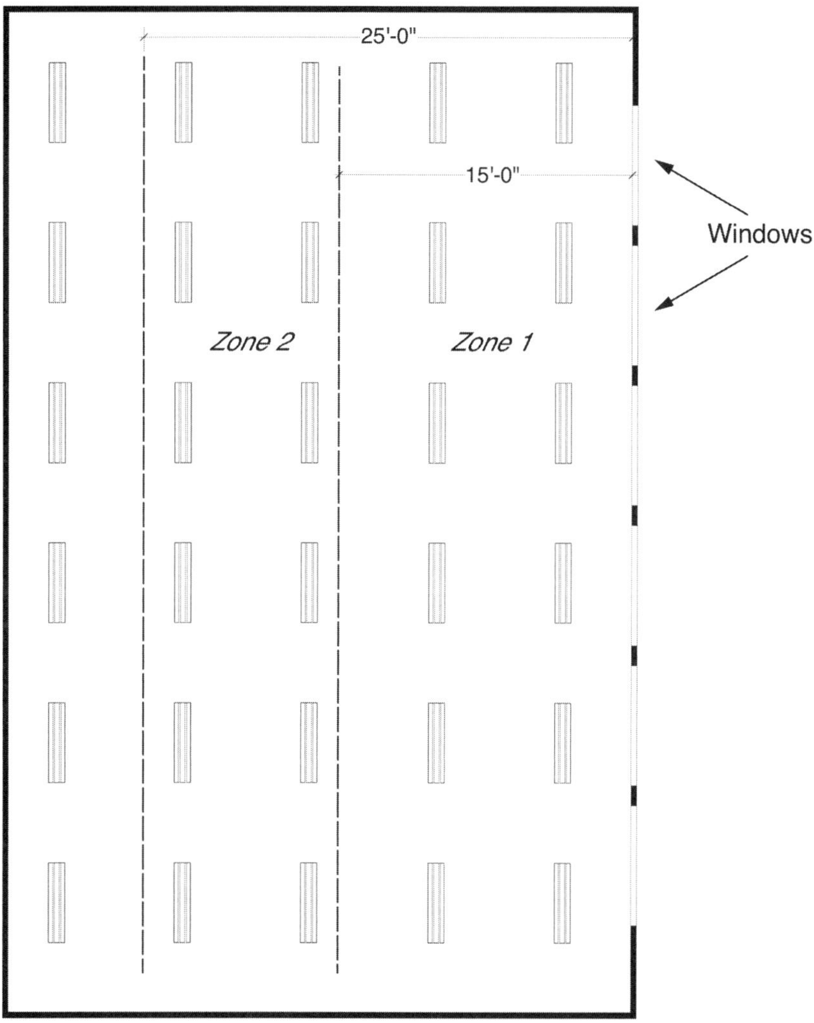


**Figure 13-1. Daylight dimming zones in a side-lit (windowed) space. The first zone includes all luminaires within 15 feet of the windows. The second zone includes luminaires in the area between 15 and 25 feet of the windows.**

of the skylight. Lighting fixtures falling within this area should be controlled together. The second zone should be calculated by adding twice the ceiling height and twice the width of the skylight. Lighting fixtures outside the first zone that fall within the area defined by the dimensions of the second zone should be controlled together. In spaces with a uniform placement of sky-

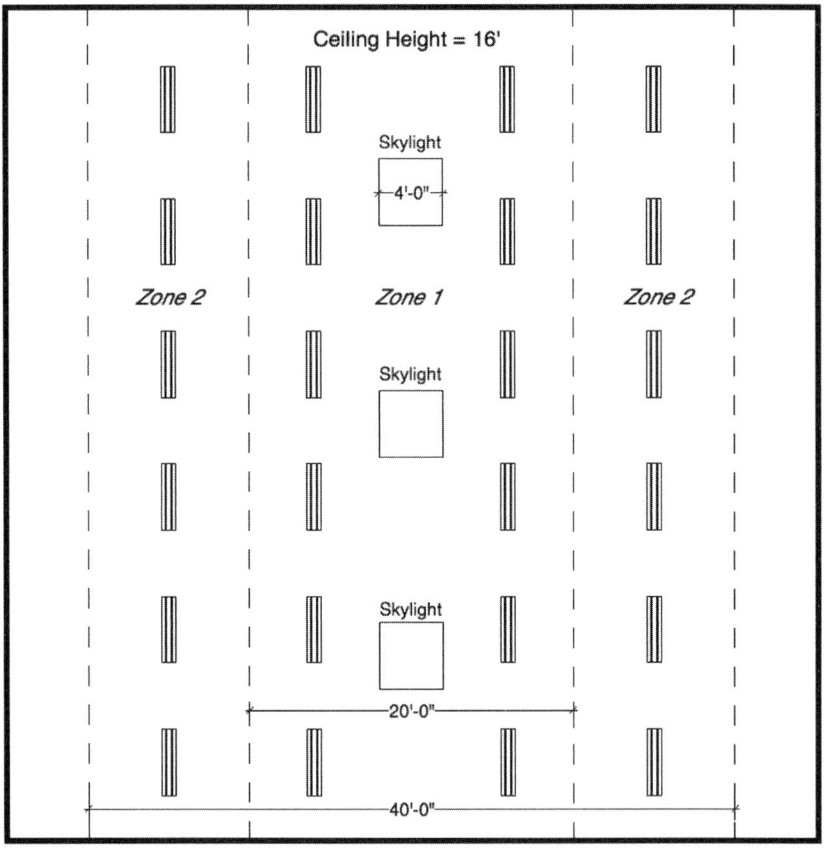


**Figure 13-2. Daylight dimming zones in a top-lit space (with skylights). The first zone includes all luminaires in the area surrounding the skylight that is defined by dimensions equal to the height of the ceiling plus the width of the skylight. The second zone includes luminaires outside the first zone that falls within the area defined by dimensions equal to double the ceiling height plus double the width of the skylight.**

lights, in which the skylight spacing is equal to or less than the ceiling height, one daylight dimming zone for the entire space is all that is usually needed (See Figure 13-2 for an illustration of these zoning parameters). These are basic parameters to consider when zoning a daylight harvesting system, and will generally apply in open plan spaces with no operable shades or movable partitions. But if a space has operable shades or other features that obstruct the daylight in one area more than another, the dimming may have to be broken down into additional zones and local controls might need to be installed so as to allow the manual adjustment of the lighting. In all cases, local building energy codes should be consulted, as these will sometimes contain requirements for the zoning of daylight dimming.

At the heart of any daylight harvesting system are devices called photosensors. These are small pieces of equipment that can be mounted in a ceiling or on a wall, in the interior or on the exterior of a building, or directly within a lighting fixture to "read" the amount of light present at any given time. By integrating photosensors into a lighting control system, and with the correct programming and commissioning, we can automatically dim or switch off the electric lighting in a particular area when the daylight contribution causes the overall levels to rise above a preset value. As previously mentioned, the "system" can be as simple as integrating a photosensor and a dimming ballast or driver in an individual lighting fixture to create a self-contained, self-dimming luminaire. Or it can be as complex as installing an array of sensors to read the light levels in a variety of spaces so as to dim each area's lighting to different levels, through a central control system, to achieve the appropriate illuminance throughout the entire building.

Daylight harvesting systems can be divided into two different basic types: open loop and closed loop. An open loop system is one in which the photosensors are positioned so as to read only the contribution of daylight in a space. They might be installed right near a window, or in a skylight, or even on the exterior of the building. Because there is no feedback, i.e. no measurement

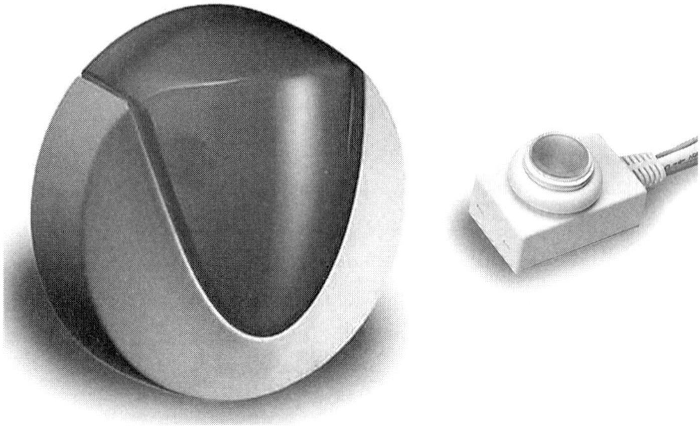

**Figure 13-3. on the left: a wall or ceiling-mounted open-loop photosensor. On the right: an integral fixture-mounted photosensor. Photos courtesy of Wattstopper.**

of the actual combined light from all sources in the space being lit, it is an open loop. A closed loop system is one in which the photosensors are positioned so as to read the combined light in the space, and it can take into account the contributions of both the electric and daylight sources, *and* any changes that are made within the space throughout the day.

Generally speaking, simpler applications can work with either open or closed loop systems, but spaces with variable lighting conditions require closed loop systems. If there are operable window shades or blinds of any kind, or if the controllable light in adjacent spaces contributes to the overall light levels in the space being controlled, then a closed loop system is the appropriate choice. Imagine, for example, a situation in which the electric lighting levels are controlled according to the input from a sensor mounted to the exterior of the building, but the windows have operable window shades. When an occupant closes the shades, perhaps to keep the direct sunlight from streaming into the room and causing glare, the system will continue to keep the lighting fixtures dimmed to the same level as when the shades were open, creating a too-dark environment. In this

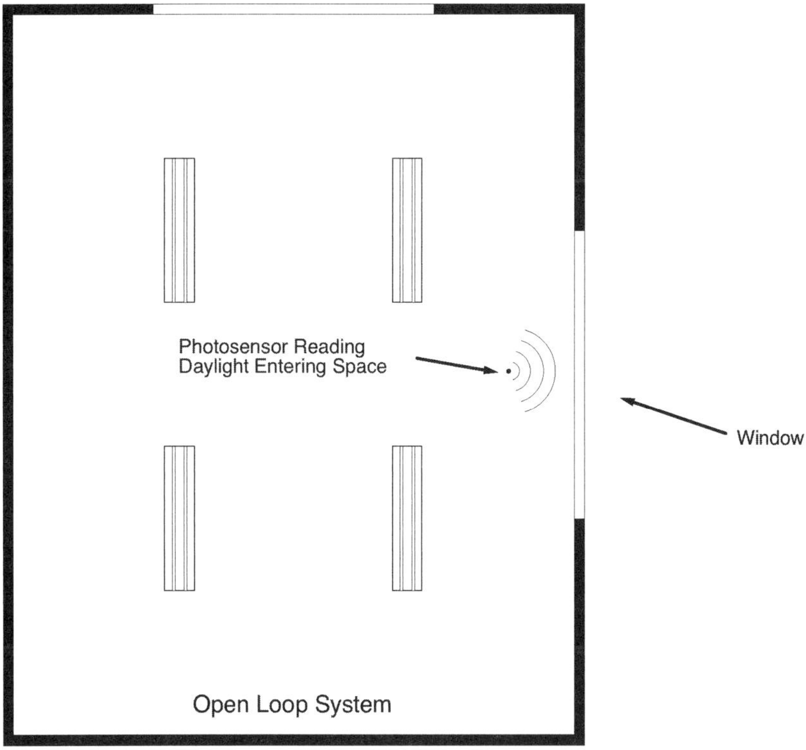


**Figure 13-4. In an open loop daylight harvesting system the daylight sensors only read the contribution of daylight to the overall light in the space.**

case, a properly zoned, closed loop system, that reads the actual amount of light in the space, will adjust the lighting accordingly when the window shades are closed. On the other hand, in a large, open plan industrial facility with regularly spaced skylights throughout and no operable shades, an open-loop system can work just fine. In general, closed loop systems function best when the photosensors are positioned so as to read areas that include work surfaces or light colored floors or walls. Dark grey carpets or other light-absorbing surfaces will not reflect an adequate amount of light for the sensors to react to in the event of a moderate change in the amount of daylight that might still noticeably reduce the illuminance in the space.

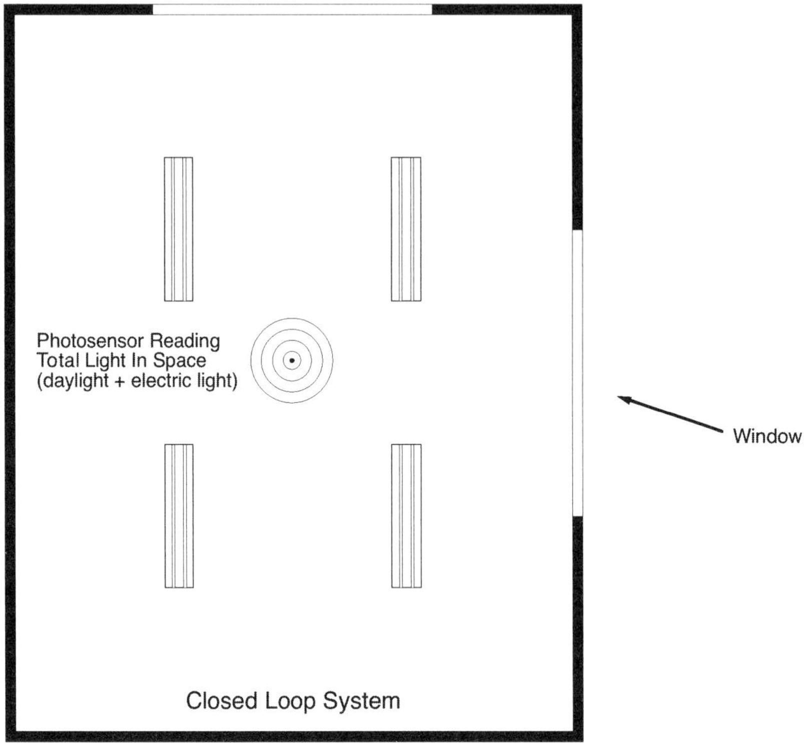


**Figure 13-5. In a closed loop daylight harvesting system, the daylight sensors read the contribution of daylight and electric lighting to the space, allowing the system to adjust the electric lighting levels when shades are drawn on a sunny day, or if there is a contribution of electric light from an adjoining space.**

Proper commissioning, the process by which we test and adjust various systems to be sure they are performing as per the design objectives and specifications, is imperative when any kind of daylight harvesting lighting controls are implemented. Set points have to be defined for the various zones, and correlated to the amount of daylight present, so that the luminaires are brought to the intensity level required to keep the overall illuminance as consistent as possible. This process takes some time, and should be performed by a trained individual. Proper commissioning may add additional cost to the project, but if it is not done correctly, the potential to save money with these systems will be seriously undermined. To complicate matters, especially in the context of today's compressed construction schedules and the often cited requirement to get the building occupied as soon as possible, it is always best to do the commissioning after all the furniture and final finishes are in place. This is because the addition of floor and wall coverings, furniture, and partitions can significantly change the performance of the lighting systems by absorbing, blocking, and reflecting the light in different ways. In addition, attention needs to be paid to keeping the system correctly programmed as the use of the space evolves and changes. When, for example, a white partition or display is temporarily set up in a retail store directly under a photosensor, the set point for that sensor may have to be changed to account for the fact that it is now going to have a significantly greater amount of light reflected back to it even though the amount of daylight, and the general illuminance levels in the room, remain unchanged. There are some "self commissioning" systems that can re-adjust themselves to accommodate changing conditions in a space. These will always be closed loop systems, but they usually cannot take the place of a commissioning technician or knowledgeable facilities maintenance staff member who will monitor the performance of the system when the space is modified. As with any system that reads an existing condition to control a component affecting that condition, the performance of a daylight dimming system is only as good as the consistency of the input. Just as when a heat source is placed too close to a thermostat after that thermostat has been set, and the space becomes uncomfortably cool, a photosensor that is not correctly set to interpolate the input it is receiving will result in a space in which the lighting levels are not maintained as per the designer's, and occupant's, requirements.

As with all sustainability strategies, we get the best result when our designs are the product of an integrated process in which the contributing elements are considered together. We can often accrue many benefits, and save energy and money, by adding daylight dimming into projects when there has been no prior consideration for special architectural daylighting treatments integral to the fabric of the building itself. But the greatest magnitude of savings will come when all these elements—lighting controls, building site and orientation, fenestration, shading, and skylights—are considered together, from the outset.

## OCCUPANCY SENSORS AND VACANCY SENSORS

Most of us grew up amid a chorus of entreaties to "turn off the lights" when leaving a room, and one of the first rules of energy conservation is to turn equipment and energy-dependent devices off when they are not in use. Occupancy and vacancy sensors are some of the simplest lighting controls to implement, and can provide energy savings by simply turning the lights off when a room is unoccupied. The difference between an occupancy and vacancy sensor is seemingly minor, but it can have a great effect on the practical results. An occupancy sensor is a motion detecting device that senses when someone enters a room, and also when that room becomes unoccupied. It can turn the lights on when the room is occupied, and then turn the lights off when it is unoccupied. A vacancy sensor is simply an occupancy sensor that is programmed to require a switch to be manually activated in order to turn the lights on, but will still turn the lights off automatically when the room is unoccupied.

When would we want to use which of these devices? To answer this question, let's first examine a couple of scenarios. From a practical point of view we would never want these devices to turn the lights off immediately upon sensing a lack of motion, as an occupant in a room can be relatively still for a few minutes at a time, and would not want the lights to go out unexpectedly. So we usually set occupancy sensors to turn the lights off between ten and fifteen minutes after the last motion is detected. As a result, an occupancy sensor will switch the lights on when anyone enters the room, and leave them on for as long a duration as has been programmed into the device, even if the occupant is just entering to quickly retrieve an item and would not have normally needed to switch on the lighting to do so. In this case, or in any other situation where someone who enters a room for just a moment would not have switched on the lights, the result can be an unnecessary expenditure of energy. And these seemingly small expenditures can add up over time. Or, if the occupant enters a daylight filled room for an extended period of time, and would not have switched the lights on at all, an occupancy sensor would switch on the lights anyway. In some cases the occupant might not bother to turn the lights off, though they were not strictly needed in the first place. With a vacancy sensor, also defined as a "manual on, auto off" occupancy sensor, the lights will remain off until the occupant manually switches them on. There are some scenarios where an occupancy sensor is the right choice. A public hallway in an apartment building and a stairwell are both examples of areas where occupancy sensors would be desirable for reasons of public safety. But vacancy sensors are the appropriate choice for many applications, like private offices and commercial break rooms, and California's Title 24, one of the United States' most stringent energy codes, mandates the use of vacancy sensors, versus occupancy sensors, in many types of spaces.

In general, vacancy sensors should always be considered for the following space types: private offices, conference rooms, break rooms, office kitchens, copy rooms, storage and stock rooms, restrooms, and any spaces that are intermittently used. Public hallways, and especially emergency stairwells, are areas where occupancy sensors, often in conjunction with bi-level luminaires, are an excellent strategy. A bi-level luminaire can be used at two different intensities: fully on and partially dimmed. In this scenario, the public hallway or stairwell, which often needs to be lit to a low illuminance level at all times to remain compliant with the local building codes, can be kept at this low level when unoccupied and then automatically lit to a higher, more comfortable level when someone enters. In the case of emergency stairwells, or stairwells in buildings also served by elevators, this is an especially good strategy as these spaces are almost always unoccupied. Occupancy sensors can also be employed in large spaces that do not need to be entirely lit all the time, like warehouses. They can be installed and zoned so as to only turn on the lights in the aisles that are occupied, and the result can be a great savings in energy costs.

A few basic types of occupancy sensors include wall mounted and ceiling mounted, in-line, and low-voltage devices. As with the photosensors we discussed in the previous chapter on daylighting, some lighting fixtures can be specified with integral occupancy sensors that are part of the fixture itself and essentially create a stand-alone occupancy sensing luminaire. Apart from these integral, fixture mounted devices, the in-line sensors are the simplest to install and use. They most often come in the form of an occupancy sensor/wall switch, which combines the function of a light switch and an occupancy sensor all in one device. These can be very easily installed in place of a light switch, often using the existing wiring, in areas where the size and shape of the room are such that the sensor will be able "see" the entire space and thereby function properly. They are "in-line" because they are essentially switching devices that are installed into the line voltage circuit that directly feeds the lighting fixtures they control. Low-voltage occupancy sensors can be wall-mounted or ceiling-mounted. They are called low-voltage because instead of interrupting the line-voltage power that feeds the luminaires, like a wall-switch, they communicate via a low-voltage digital or analog signal with the luminaires, or with another in-line control device that might be a relay switch, a series of dimming ballasts, or a complex, centralized control system, to "tell" the luminaires what to do. (A relay is a remote-controlled switch that controls a high-voltage device, activated by a low-voltage signal from its remote counterpart.) Low voltage occupancy sensors are most often used in more complex control scenarios, where the lighting being controlled runs across multiple control zones, or in areas where multiple sensors are required because of physical obstructions or irregular room geometries.

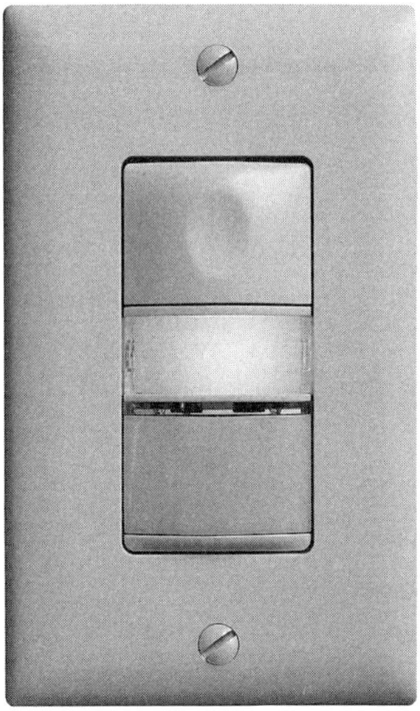

**Figure 13-6. An in-line wall switch occupancy sensor. These simple wall switch replacements are extremely easy to install and use, and can provide effective occupancy sensing in regularly shaped, small to medium sized rooms. Photo courtesy of Wattstopper.**

There are two types of technologies used in today's occupancy sensors, *passive infrared* and *ultrasonic*, which are more or less accurate in different situations. Dual technology sensors utilize both of these in one sensor to produce more reliable results in many applications. **Passive infrared** (PIR) sensors are sensitive to heat, including the heat generated by a human occupant, and can detect the movement of a person in a direct line of sight to the sensor. PIR sensors cannot "see" through obstructions, like walls or glass, or around corners. In addition, PIR sensors will detect motion across their field of view, rather than towards or away from the sensor. False triggering can be caused by people walking by open doorways or windows, or sunlight and HVAC outlets that periodically heat up objects in the room. **Ultrasonic** sensors work by emitting sound waves outside the range detectable by human ears and then "hearing" the reflected sound when it returns back from the objects and surfaces in the room, much the way a bat "sees" in the dark. They are good at detecting motion toward and away from the sensor, rather than across the space. Ultrasonic sensors can "see" through obstructions, like restroom partitions, and around corners. They can be falsely triggered by people walking in adjacent spaces within the sensor's range, and by objects blown around by the wind or a ventilation system. Dual technology occupancy sensors combine both passive infrared and ultrasonic technologies in one sensor, and are the best at avoiding false triggering as they weigh input from both protocols before turning on the lights. They are also more expensive. A study by the Lighting Research Center at the Rensselaer Polytechnic Institute in Troy, NY has shown that occupancy sensors, when correctly specified and placed, can save between 30 and 40% of the energy used to light a shared space, and 25 percent of the energy used to light an "owned" space. [15] Owned spaces are those with a primary occupant, like an office or a single-teacher classroom. The savings are less for owned spaces because occupants who are the sole, or primary "owners" of a space are more likely to turn off the lights themselves when they leave.

**Fluorescent Lamps and Occupancy Sensors**

An important consideration when using fluorescent luminaires in conjunction with occupancy sensors is lamp life. Fluorescent lamps are rated with a lifetime defined by a certain number of hours of operation that is also contingent upon the lamp being "started" a set number of times. Some lamp manufacturers even give their lamps two lamp life ratings: one that assumes a twelve-hour start (i.e. with the lamp started once every twelve hours of operation) and one that assumes a three-hour start. The more frequently a fluorescent lamp is started, the shorter its life will be. Typically, when a fluorescent lamp is started, a surge of high voltage current is momentarily applied to the lamp to ignite it, which results in some additional wear to the lamp electrodes. Instant start ballasts work in this manner. They are the most common electronic ballast used today, and also the most energy-efficient in terms of the efficacy (lumens per watt) when used with the proper lamp. Another ballast type designed to mitigate this wear and tear to the lamp by preheating the electrodes is the rapid start ballast. But using these ballasts will result in slightly decreased energy efficiency as they will continue to apply pre-heat or "glow" current to the lamp after it has started and for the entire time it is on. Program start ballasts are essentially more advanced versions of the rapid start ballast that precisely time the preheating of the lamp electrodes before applying the voltage required to ignite the lamp, resulting in the longest lamp life in scenarios where the lamp is switched on and off many times a day. When we use occupancy sensors in conjunction with fluorescent lighting, or in any situation where we can expect fluorescent lighting to be switched on and off frequently, program start ballasts should be specified. From an energy efficiency standpoint, the additional energy used to run the lamps will be more than offset by the energy saved on having the lights turned off in sporadically occupied spaces. And the increased lamp life will certainly save energy, and money, in the form of fewer lamps produced, shipped, and eventually discarded.

## TIME-OF-DAY CONTROLS

Time-of-day controls, or scheduling, can be a very important, and effective part of an energy-efficient lighting control system. They automatically turn the lights off, or reduce their intensity, to reflect the changing requirements within a space from day to night, or as working hours begin and end. These systems are commonly used to control exterior lighting and shared interior areas of commercial and public buildings where there is no single "owner" of the space (i.e. where there is no one who will reliably be the first in or last out each day and who can be tasked with turning the lights on and off at the appropriate times). Lighting in open offices, public hallways, and other circulation spaces are all prime candidates for these types of controls, as are exterior areas, facades, and landscape lighting. Time-of-day controls will automatically turn the lights on at a certain time of the day (or night) and turn them off at others. But what is meant by "time-of-day"? There are applications where we might want the lights to go on and off at the same clock-time—seven in the morning and six in the evening, for example—like an office space where we come to work and go home at the same time each day. But there are also applications, like exterior area and facade lighting, where we may want the schedule to be timed to coincide with the actual sunrise and sunset, which changes with the seasons.

And instead of simply switching the lighting fixtures on or off, in some cases we might prefer to dim them, like a corporate installation where the circulation areas and hallways should be lit for security reasons and for the benefit of after-hours workers, but don't need to be lit to a full level all through the night.

Depending on the needs of the project, there are a wide variety of time-of-day control systems available. These range from simple, mechanical clocks that open or close a few relays to switch one set of lights on and off at the same time each day, to computerized systems that can switch or dim dozens of zones to different levels. Most of the latter type utilize astronomical clocks that automatically change the clock time of their programs a little bit each day, so as to follow the actual time of sunup and sundown. The more complex, computerized controls have traditionally been reserved for larger systems comprising sixteen zones or more, but increasing numbers of manufacturers are now incorporating the same functionality in smaller, more cost-effective, modular panels, that are designed to control as few as four zones but can be networked together to create larger-scale systems. The additional functionality of these systems include the ability to control dimming as well as switching, and they can be integrated with daylight harvesting systems, bi-level occupancy controls, and local, manually-operated switching and dimming controls. These more complex systems will require commissioning, and can be more complicated to program. On the other hand, the simpler time clock and contactor relay systems, in addition to offering many fewer options, may require new wiring and conduit, an expensive prospect, whenever the zoning or scheduling needs to be changed.

An important consideration when using time-of-day control systems is the integration of manual or automatic overrides, so that occupants in a given space outside of the normally programmed working hours will not be left in the dark. These can take the form of a manual wall switch in a private office or an occupancy sensor in a public area or open office. In either case, the use of occupancy sensors may be advised to insure the lighting is turned off again when the space is unoccupied after an occupant manually overrides the schedule instituted by the time-of-day controls.

## LOAD SHEDDING

Load shedding refers to a variety of strategies to reduce the overall load on the electrical power grid during times of peak demand, when the utility companies and the grid are most stressed. As our energy use has increased over the years, improvements to the power grid—the system of high-voltage wires that distributes electricity to our homes and businesses—have not kept up, adding to this stress. Peak demand usually occurs during the day, and often in the summer, when most people are awake and at work and air conditioning systems are in use by a majority of commercial and residential consumers. Because of the added cost to build power plants with the capacity to generate more power than is usually required (and can usually be sold), it is this peak demand that is the most expensive to satisfy. As a result, utility companies often charge more for each kilowatt-hour used by its commercial customers during peak demand hours. (A kilowatt-hour, the standard measure of electrical energy for the purpose of billing charges to the consumer, is the amount of power that is required to support a 1,000 watt load for an hour of continuous use.) Some power companies build their plants with sufficient capacity to satisfy peak demand with the knowledge that they can charge more for this additional usage when it occurs. And others simply buy the extra power from other utilities when their customers need it, with the result that the electricity is sometimes generated hundreds, or even over a thousand miles away, especially during times of peak demand. Managing this electrical traffic is a complicated task, and a breakdown of any component of the system—a failed power plant, or a sagging transmission line—can result in an overwhelming degree of stress to the grid. Add to this the fact that higher temperatures during a hot summer day, when demand is likely to be greatest, can contribute to electrical system failures; and the result is a somewhat fragile system in which the current economic models do not support the infrastructure required to reliably provide enough power during these times of peak demand. In August, 2003, a sequence of events, in which a power plant went off-line in Ohio on a very hot day, caused the overloading and overheating of power lines as a surge of electricity from other plants flowed into the grid to make up for the lost power from the failed generators. The result was that a large number of power plants went off-line, and approximately 55 million people in eight U.S. states and Ontario, Canada, including almost all of New York State, lost power. The outages lasted from a few hours to over a day in some locations. To be sure, this is an extreme example in which a perfect storm of events occurred to create a large-scale failure. But it is also illustrative of the fragility of our electrical system at a time when a new economic reality, coupled with a lack of political will to make investments in public infrastructure, amid ever-increasing energy use, is creating a widening gap between our demand for energy and the readily available supply at our disposal. Load shedding can be an effective strategy to manage this gap.

Load shedding, which we can also think of as energy-use reduction to satisfy a temporary scarcity of resources, has traditionally been applied via strategies to reduce the use of HVAC systems at times of peak demand. These strategies have taken many forms, from published notices the night before requesting that building owners and operators change their air conditioning set-points the next day because of a forecasted high demand, to systems that automatically send a request from the utility to the customer as the demand peaks, asking them to make these kinds of changes instantaneously. But studies by the Rensselaer Polytechnic Institute's Lighting Research Center suggest that lighting systems can also provide a very effective strategy for load shedding by dimming the lighting to reduce energy use during the times of peak demand. Whereas a small change in temperature and humidity can make some people uncomfortable when air conditioning set points are raised, these studies have shown that reducing the intensity of interior lighting is well tolerated among a majority of people. Reducing the illuminance by as much as 15% will often go unnoticed when it is done gradually, and reducing it by a factor of 30% or more will be acceptable to most people when it is coupled with education that explains the reason for the reduction along with assurances that the change is temporary. [16] In many ways, daylight dimming is a natural load shedding technique, as it automatically dims lighting during the day, in daylit areas, when peak demand is at its greatest. But we can also incorporate load shedding in non-daylit areas by implementing programs and technology that automatically dim the lighting in response to a request from the utility company. One way this can be accomplished is with the use of inexpensive bi-level or step-dimming ballasts for fluorescent lighting that respond to a signal sent over the power lines. This is a very cost-effective solution, especially suited to retrofit applications, as it requires no additional control wiring to be added to the building, and the ballasts are less expensive than continuous-dimming ballasts. (We will discuss step-dimming and continuous-dimming in greater detail in the next section of this chapter). But this kind of system has its disadvantages, as the sudden change in lighting levels that results with step-dimming can be distracting, and is generally less desirable than the gradual change produced by a continuous dimming system. Continuous dimming can also be used to automatically reduce the lighting load at times of peak demand. Either way, incorporating load shedding into a lighting control system is a good strategy to minimize stress on the electrical infrastructure, and to reduce energy use. And since we are reducing usage when the energy is at its most expensive, the payback for these systems can be more easily made up over a shorter period of time.

## CONTINUOUS DIMMING AND STEP DIMMING

Continuous dimming and step dimming are different methods of control that will bring the lighting in a space or area to a lower level. Both can enable the occupants of a space to adjust the lighting to suit their individual preferences, whether it's via a wall-mounted or luminaire-mounted switch or dimmer control. These same methods of dimming are also used to adjust the intensity of the lighting automatically, by allowing a control system to dim the lighting in response to the presence of daylight (daylight harvesting), the absence of any occupants (occupant sensing), or as a means to reduce the electrical load of the system at a time of peak demand (load shedding). In these cases the control is triggered via input from a photosensor, an occupancy sensor, or a load shedding request from the utility. Most of us are familiar with continuous dimming, as this is the kind of dimming control that is likely to be installed in our homes, shops, restaurants, and theaters, where we have regularly seen dimming controls utilized for many years. Step-dimming is a less nuanced method of control which has its place as an inexpensive energy-saving control strategy, especially in some industrial and commercial applications and when we are controlling the lighting for reasons of public safety, as in emergency stairwells and some public corridors.

Step dimming, or bi-level switching, reduces the output of the source by a set percentage. It results in an abrupt change of the lighting level that can be distracting, but is acceptable in many situations. Continuous dimming allows us to smoothly lower the level of the light source from full to a very low intensity, before it goes out entirely. Continuous dimming is recommended for any situation where we want to precisely choose the level of lighting, or where the intent is to bring the lights to a different level gradually, so as to make a change that is imperceptible or barely noticed. Daylight dimming and load shedding are examples of scenarios where continuous dimming will often be preferred, as the intention is to lower the electric lighting level in response to an increase in daylight, or at a time of peak demand, without distracting the occupants of the space. In addition, by allowing for a more precise level of control over the lighting, continuous dimming is a far more effective way to accommodate individual preferences. It is often the case that different groups in an organization—photo editors and writers at a newspaper, for example—will desire different amounts of light to do their jobs. This can be because they are working in different mediums, on paper or with a computer, or because the tasks they are engaged in involve larger or smaller objects and therefore require different degrees of visual acuity. Office workers in general have demonstrated a great preference to be able to adjust the level of lighting in their workspaces, and when given the opportunity they have more often than not chosen settings that resulted in illuminance levels lower than the recommended standards. It may not be wise to extrapolate from these findings that lower illuminance levels should be recommended across the board, due to the fact that a small number of subjects chose levels higher than the current recommendations. But an effective energy-saving approach might be to provide for lowered illuminances throughout shared workspaces while simultaneously providing individual or group controls to adjust the lighting levels in different areas up or down. It has been estimated that providing for these individual preferences by giving office workers continuous dimming control over their lighting, would result in an overall energy savings of 20 to 40% for the controlled lighting. [17] But continuous dimming systems can be expensive, largely due to the added cost of the dimming ballasts required for every fluorescent luminaire. Dimming drivers for LED luminaires are less expensive than dimming ballasts for fluorescents, and we can expect the cost of continuous dimming systems to decrease as LED lighting technology matures and begins to replace fluorescent lighting as the most cost-effective option. It's also important to consider that while energy use will go down when lighting is dimmed, the overall efficacy will go down as well when using fluorescent dimming ballasts, offsetting this savings to some degree. With LED lighting the efficacy will remain the same, or possibly even improve, as the source is dimmed.

As noted in the chapter on occupancy sensors, step dimming, or bi-level switching can be very effective when used in certain applications. Stairwells and public hallways that need to be lit to a certain level at all times to satisfy the building code are one example. With a step dimming system we can program stairwell lighting to run continuously at a very low level and then automatically come to full when someone enters the space and triggers an occupancy sensor. Large warehouses in which it is unnecessary to have the light at a full level in the unoccupied aisles, but it is also uncomfortable for the workers to leave large parts of the space entirely dark, are another example of an application suited to step dimming controlled by occupancy sensors. In this case bi-level luminaires might be left on at a low level throughout the warehouse, with occupancy sensors arrayed so as to bring the luminaires in individual aisles up to a higher level when a worker approaches. Step dimming can be as simple as having half the lamps in the fluorescent fixtures in an entire space controlled by one wall-switch with the other half controlled by another, and some fluorescent ballasts are manufactured with this functionality built right in. In a classroom setting the fluorescent lighting might be of this type, allowing the lighting level to be reduced by one-half with the simple flick of a switch when projected instructional material is used, without the considerable expense of incorporating dimming ballasts in all of the lighting fixtures.

Another factor that must be considered when choosing a dimming system, be it step dimming or continuous dimming, is the source employed within the luminaires to be dimmed. As we will discuss in greater detail in the next section of this book, the way dimming works can be quite different for different types of sources, and dimming generally works much better with some than with others.

## LIGHTING CONTROLS, BALLASTS, DRIVERS, AND SOURCES

In order to specify a lighting control system we first need to develop an understanding of the programmatic and visual requirements of the occupants who will be working, or living, in the spaces where the controls are implemented. Will dimming be a requirement or is bi-level or multi-level switching an appropriate solution? Is it necessary to be able to dim the lighting to very low levels? If so, we may need to consider the external factors affecting the space (like daylight infiltration or light that may spill over from adjoining spaces). Equally important is a clear understanding of the capabilities and functionality of the different sources we are employing. Not all sources lend themselves to all types of controls, and not all sources will work well in all of the possible scenarios where controls are likely to be specified. For example, LED, incandescent, halogen, and fluorescent sources are dimmable to very low levels, and are therefore good candidates for applications where smooth, continuous dimming is desired. But not all LED or fluorescent luminaires dim equally well with all of the different kinds of control systems that are available. HID and magnetic induction sources cannot be dimmed to very low levels at all, and continuous dimming of these is generally not recommended. When dimmed, HIDs will often produce a light with reduced color rendition. In addition, the dimming of HIDs can be problematic on a practical level. Many of these systems require the lamp to be brought to full output for a period of minutes before they can be dimmed, and the stability of the lamp can decrease at lower levels, resulting in a noticeable flicker. Step dimming or bi-level switching may still be a good solution for controlling HIDs in parking garages or warehouses, but it is probably just a matter of time before fluorescent and LED solutions completely replace it in these kinds of applications as well. In addition, HID lamps cannot be immediately restarted after they are switched off, requiring instead a cool-down period of a few minutes. And when HIDs are switched on they go through an unattractive startup cycle in which they exhibit very noticeable color shifts before coming to full intensity. This makes HIDs unsuitable for use where the lighting will be switched on and off frequently, as is the case in an intermittently occupied space with occupancy sensors programmed to completely shut off the lighting, though, as mentioned above, there are bi-level HID luminaires that can be used in these scenarios. Clearly then, HID and induction sources are equally unsuited to applications where the lighting needs to be dimmed smoothly, to very low levels, as is the case in a corporate presentation room or a theater, whereas LED, fluorescent, and halogen sources are good choices when it is important to have this degree of control.

Once we have established that a particular source is suitable for a certain control scenario we then have to identify the kind of system that works with that source, and will provide the functionality required by the end users. In addition, we need to consider if it is a good fit for the building project. Some control systems are better suited to retrofit projects, and some are more easily integrated into new construction. There are many kinds of lighting control systems and devices of varying degrees of complexity, and often more than one that will work for a given application. There are very simple to install, in-line, wall-box switches, dimmers, and occupancy sensors that provide local control in a single room or area; equally simple to install standalone systems comprised of arrays of individual luminaires with integrated photosensors and occupancy sensors that essentially control themselves; and complex, centralized systems with relay switches and dimmers located in remote electrical closets that regulate the power to circuits distributed throughout a facility. It might be that local control via individual switches and dimmers on a wall with integral occupancy sensors are all that's needed. Or maybe the programmatic requirements of the users call for a more-complex, facility-wide control system with manually activated override buttons and an automatic time-clock that can be programmed to recall a number of pre-set scenes, each with different control zones brought to different pre-set levels. (Control zone is a term we use to describe a group of luminaires, usually of the same type and application, comprising a discrete treatment, and that are controlled together. An example of a control zone might be the wall-washers on the south side of a lobby.) Or, as is often the case nowadays, we might have a modular, distributed control system that consists of a series of input devices—local control switches, dimmer controls, photosensors, occupancy sensors, and an astronomical time clock—that send commands to a digital network of relays and dimmers distributed throughout a facility in order to control the lighting in each space from anywhere—including a remote location via an internet connection. In some cases these input devices might send digital commands directly to luminaires fitted with dimming drivers and ballasts that can dim or switch themselves. And in many of these situations the question can become, "Where does the control system end and the luminaire begin, and is it always so easy to separate the two?"

Our traditional dimming systems, still used in smaller retail environments and homes, are based on the technology and electrical infrastructure required to control incandescent and halogen lighting. If you recall our earlier discussion of sources, both of these types do not require a ballast, but instead run off the line voltage that typically serves our buildings or, in the case of some halogen lighting, low voltage via a transformer feeding off the line voltage circuits. Both can be dimmed very easily by directly regulating the power feeding the luminaire itself. This can be done via an in-line wall-box dimmer that is installed in place of a wall switch, or, in a larger facility, through the use of a centralized dimmer panel in an electrical closet that responds to a signal or command from a wall-mounted control device, a sensor, or an automatic clock. In both of these scenarios, we are dimming the luminaires by directly regulating the line-voltage that powers them through the use of a standard incandescent or phase control dimmer. But the rest of our commonly used sources and lamps require the integration of a device—a ballast or driver—to regulate the electrical current that starts and runs them. This device is usually incorporated within the body of the luminaire, whether it's a ballast for fluorescent or HID lamps or a driver for LED luminaires, though it can also be located remotely (in the ceiling plenum or a wall, for example) in cases where the luminaire design is not large enough to accommodate it. It's these ballasts and drivers that actually dim, or control, the light output of the source, and thereby the luminaire, when activated by the control system. Though these ballasts and drivers are considered a part of the luminaire, and are specific to the light source, they are undeniably also a key component of the overall control system.

## LOW-VOLTAGE CONTROL SYSTEMS & PROTOCOLS

Clearly then, ballasts and drivers constitute an important link in the control chain that we need to consider and specify along with the front-end input devices and control systems. Not all ballasts and drivers support dimming of the sources they control, and the ones that do may function in a number of different ways that require different control inputs. Some fluorescent ballasts and LED drivers are designed to dim their sources in response to a typical incandescent or low-voltage dimmer that regulates the line-voltage current serving the luminaire, making them compatible with an existing building's electrical infrastructure, and therefore easy to install. This approach creates products that can be easily retrofitted into existing buildings, and that will work out of the box with older control systems, but it will not generally result in the best performance from the luminaires, or be the most sustainable design solution in the case of a new construction project. All the better-performing fluorescent ballasts and LED drivers now being manufactured that accommodate dimming are designed to respond to one or another of the variety of control protocols, or languages, used by low-voltage control systems. These systems are referred to as "low-voltage" because they rely on a signal current, usually lower in voltage than the line-voltage current that actually powers the light sources, to dim the luminaires they control. Each of these different control protocols will require a specific wiring scheme, or topology, to function. And they each have their own set of advantages and drawbacks.

### Analog Controls

A very common, tried and true control protocol is the analog, four wire, zero-to-ten-volt (0-10v) system that has been used to dim a good deal of the fluorescent lighting in commercial installations for some time now. A 0-10v system uses a separate, low voltage control circuit in addition to the high-voltage circuit that powers the luminaires. This low-voltage circuit runs a "signal" current to "tell" the ballast or driver what level to drive the source. In this system, the signal voltage theoretically relates to the percentage the lamps are to be driven. When the system is sending 10 volts over the control wire, the ballasts will drive the lamps at 100% of their intensity. When the system is sending 2 volts over the control wire, the ballasts will drive the lamps to 20% of their intensity. 0-10v systems, including luminaires fitted with compatible ballasts and drivers, are readily available and are easily understood by electrical contractors and manufacturers alike. But from a sustainability perspective, the requirement to run a complete low-voltage circuit as well as a dedicated high-voltage circuit for each control zone necessitates the use of a good deal more in the way of materials and resources (i.e., copper wire) than the newer, digital systems, which we will discuss later in this chapter. In addition, the labor costs to run the additional wires can add up, making these traditional systems costly to install. On the other hand, in the case of a simple, localized application where centralized control of an entire facility is not required, the 0-10v protocol can be a cost-effective way to allow the lighting in a single space to be dimmed. This can be accomplished through the manual activation of a wall dimmer, or via the automatic input of an occupancy sensor, or with a photosensor that detects the presence of daylight. In addition, the 0-10v protocol remains a reliable common denominator whereby any of the digital control systems currently on the market can all be used to dim any luminaire that can be specified with a 0-10v dimming driver or ballast. This is usually accomplished via the integration of a device in close proximity to the luminaire that translates the digital protocol to a 0-10v signal. Most of the manufacturers of digital control systems will offer these devices so as to allow their systems to be used with a wider variety of luminaires.

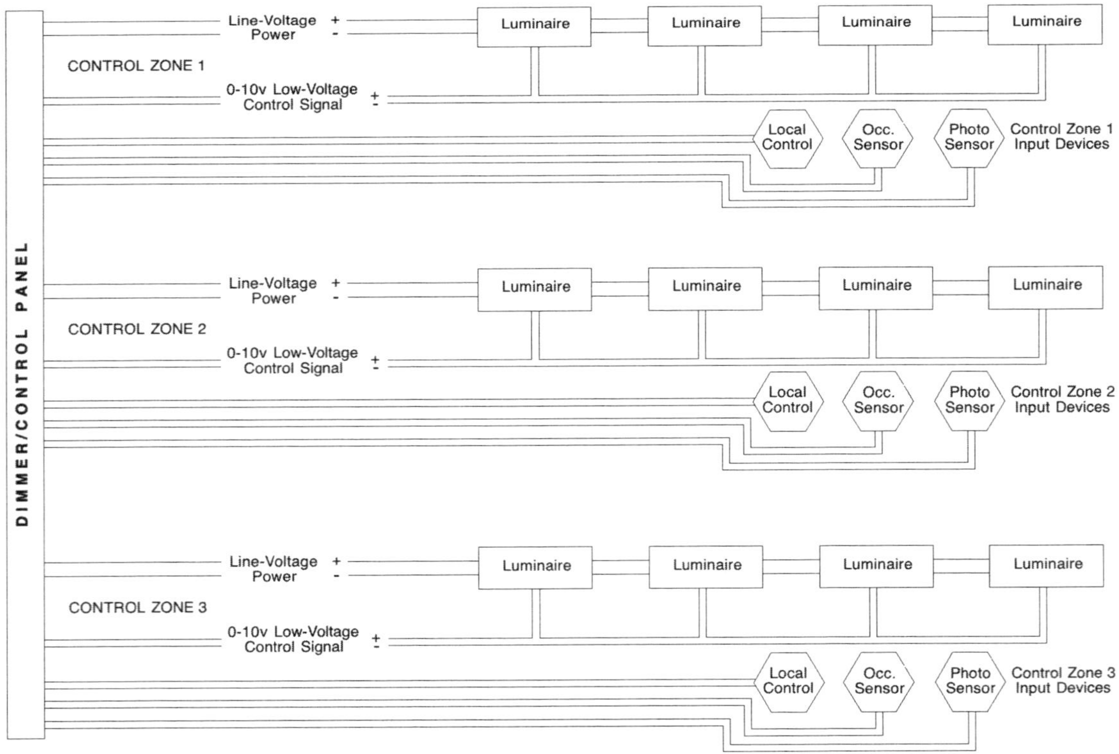


**Figure 13-7. Wiring diagram for a centralized, analog 0-10v control system. This system requires the use of four dedicated conductors per control zone, two for the line-voltage luminaire power, and two for the low-voltage control signal, as well as separate wiring for the input devices (local control devices & sensors).**

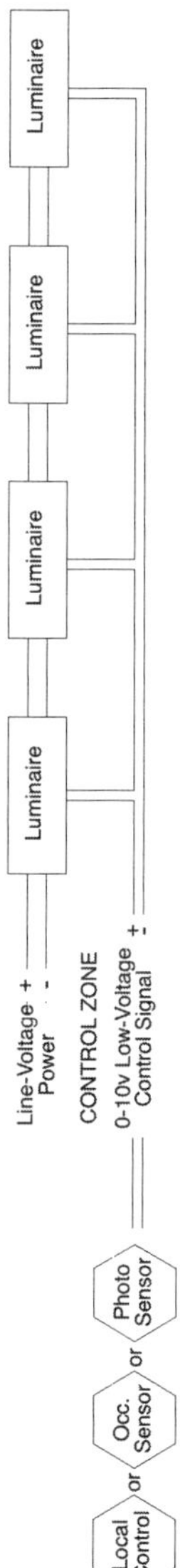


**Figure 13-8. Wiring diagram for a simple, single zone, 0-10v control.**

Another analog protocol used to control fluorescent lighting, and now available for LED lighting, is a line-voltage, three-wire dimming system. This system makes use of two wires carrying hot (positive) line-voltage currents—one at a constant voltage and one that is variable—along with one neutral (negative) return wire. The three-wire system is similar to the 0-10v system in that a constant voltage is supplied to the ballast via one wire to actually drive the luminaire, and a variable signal current is supplied with another wire to control the level of dimming. But with the three-wire system the signal current is transmitted via the second, high-voltage conductor instead of a separate, low voltage circuit. This system is easy to wire, and uses one fewer conductor than the 0-10v system. But each control zone still needs a dedicated run of all three conductors from the luminaires in that control zone to the input device.

**Digital, Distributed Controls**

Digital, distributed controls are the newest and most flexible systems for lighting control. They are the easiest to install, though they represent a sea change in the way a control system is laid out, and may pose a challenge for some electrical contractors who are unfamiliar with them. Their main advantage, from the perspective of an installer, is that they require fewer individual runs of wire and therefore provide for potential savings in both labor costs and materials. With digital controls we can literally create a network of an entire building's ballasts, drivers, dimmers, relays, and input devices, assigning a unique digital address to each, so as to allow any control device on the system to control any individual lighting fixture alone, or in an infinite number of user-definable groups. With these systems the control wiring is non-linear, which means it can be run in the most convenient way possible with little regard as to which device is directly connected to which. A major benefit of this kind of system is that it can be easily reconfigured as the use of a space changes, and even as partitions and walls are added and removed, without requiring any rewiring to modify the control

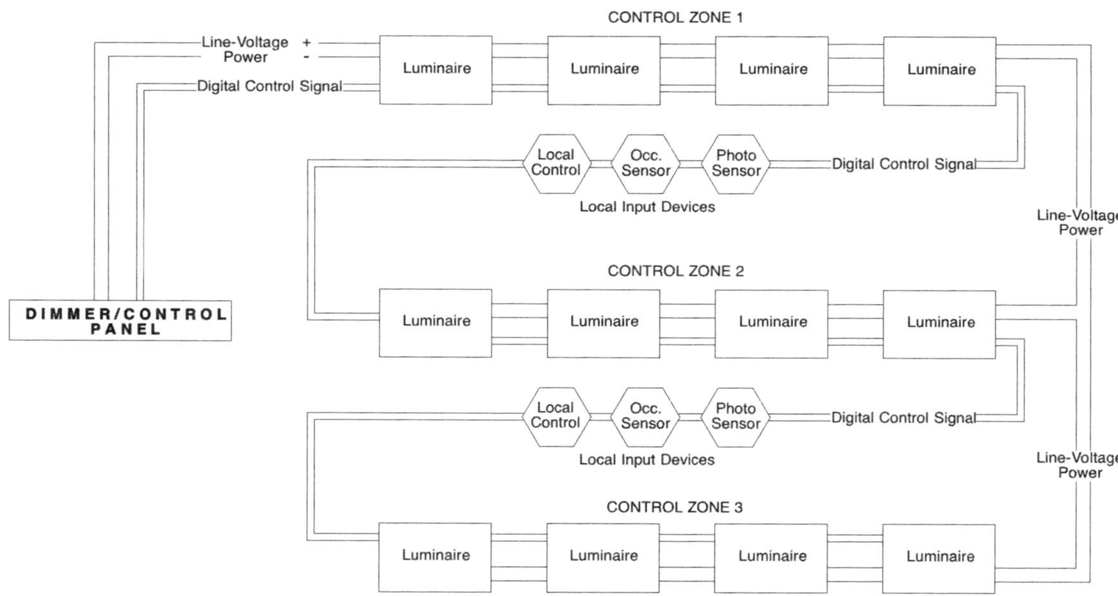


**Figure 13-9. Sample wiring diagram for a distributed, digital control system. Luminaires from multiple control zones can be powered from the same line-voltage circuit, and they can share the same control wire loop with input devices in different areas or spaces. Luminaires and control devices are all digitally addressable, and all control zones, and the input devices they will respond to, are defined via programming and can be easily changed when the space use changes.**

zones so they conform to the new space layout. Unfortunately, there is little standardization in the marketplace with regard to these systems, and as a result each employs a different digital protocol, or language, for communication between the control system and the luminaires, making it difficult to source different components from different manufacturers.

DMX512 (*Digital Multiplex with 512 pieces of information*), the first digital system for lighting control to be commercially available, was developed in the late 1980's by leaders in the theatrical and entertainment lighting industry as an answer to the ever-increasing complexity of the lighting systems they were employing. Previously, these lighting systems were most often run via 0-10v control systems that became ungainly as the low-voltage wires required for the large number of dimmers and devices in a single show began to multiply exponentially. In response, some manufacturers of dimmers and controllers developed their own proprietary digital systems in which the communication between dimmers and devices was facilitated via a two-conductor control wire that ran from device to device. But as the number of devices requiring control continued to grow to include larger numbers of dimmers, automated lighting fixtures, and accessory equipment, all by different manufacturers, the need to develop a standard whereby all of these different devices could talk to one another, and to one control system, became clear. The DMX standard is a very robust protocol that supports the simultaneous control of multiple devices by multiple devices, allowing multiple controllers to function in concert and send multiple commands to a large number of dimmers, fixtures, and accessory devices, all with different control parameters and timings.

DALI *(Digital Addressable Lighting Interface)* is another digital lighting control standard developed by a group of fluorescent ballast manufacturers in the mid 1990's and made commercially available by the end of that decade. The DALI protocol, designed as it was for the fluorescent lighting market, provides a somewhat less powerful method of control than DMX512, which was developed for the far more complex systems that are common in the entertainment industry. DALI is slower than DMX, making it unsuited for the control of color changing LEDs and other special theatrical effects, and it is limited in terms of its ability to simultaneously talk to multiple devices. But some manufacturers have successfully adapted DALI into proprietary control protocols that allow for greater functionality in this regard, and as a result there are DALI-based systems available that will seamlessly integrate commands from multiple input devices to multiple groups of luminaires. This is often a requirement in the case of a complex, multi-tiered, control system whereby different groups of luminaires might be simultaneously controlled via inputs from manual controls and a daylight sensor, or a daylight sensor and an occupancy sensor, or any other combination of input devices. The DALI standard is not as tightly written a specification as DMX512, leaving more room for interpretation by the manufacturers who make equipment utilizing this protocol, with the result that control devices by one manufacturer may or may not always work perfectly with ballasts and drivers from another. Care must be taken when specifying a DALI system to use components from manufacturers with a proven track record of compatibility. DALI does have the added advantage of being bi-directional, enabling lighting fixtures to "talk back" to the system and report on their status, thereby alerting the operator as to whether they are on, off, or in need of maintenance.

With no adopted standard and many proprietary (and some open source) systems available, the somewhat fractured state of the architectural lighting controls industry with regard to digital controls makes the specification of these systems more difficult. But these systems hold great promise for increased energy efficiencies, materials conservation, and reduced construction costs. The ability to link thousands of devices together on a network and control them as user-definable groups, with no regard to the physical topology of the actual wiring, allows the programming of complex control scenarios that can be continually adapted as the use of the spaces within a building changes, even as those spaces are subdivided or combined. With analog controls, the physical wiring of a control zone for the lighting in a particular space has to be re-done if the use of the space changes, or if the space is divided. This can be a costly procedure, and one that might be overlooked when the space is altered by a new occupant, with the result that control systems designed to limit energy use in the old space are rendered ineffective, and must be subsequently removed from service.

To be sure, the successful incorporation of digital controls in a complex scenario requires thought and consideration in order to specify control devices, relays, and luminaires that speak the same language. Whatever system is employed, the driver or ballast in the luminaire must be compatible with the digital "language" of that particular control system. Specifying compatible luminaires and controls is an important task for the lighting designer, and one that can be daunting in a large project, especially when working with a controls or luminaire manufacturer whose product literature is incomplete. When the control options for a particular luminaire are limited (as is often the case) a hybrid digital/analog system can be employed that utilizes a digital controls system with analog 0-10v relays or dimming devices that are installed in close proximity to luminaires fitted with 0-10v dimming ballasts or drivers. This strategy can have the advantage of allowing the specifier to choose from the broad range of luminaires on the market and incorporate them into any control system. But depending on how it is implemented, this tactic may not deliver on the full potential of a true digital system. It should also be noted that digital systems require some time after installation for proper commissioning and programming. And their programs must be maintained by a facilities manager or maintenance staff member who understands the system, and how to adapt it when the space use, or the requirements of the occupants, changes. If these systems are not programmed at the outset so as to properly serve the requirements of the occupants, and if they are not maintained as these requirements change, they may cease to function as originally intended, and any potential energy savings may be lost.

## WIRELESS CONTROLS

Wireless controls can be a cost-effective solution for retrofit projects as well as new lighting systems in existing buildings. These systems are comprised of dimmers (often 0-10v) and relays with built-in wireless receivers, activated by controls and input devices that transmit a wireless control signal. This allows for the incorporation of advanced controls without the expense of adding additional conduit and wiring into the ceiling and walls in existing buildings, greatly reducing the cost to install them. With these systems, the dimmers and relays that actually control the luminaires are installed in close physical proximity to them, in an adjoining junction box in the ceiling for example, or in place of an existing wall switch. Occupancy sensors, photosensors, or additional switches, can then be installed anywhere else within range, with no need for a physical connection between the two. To simplify things further, there are self-powered wireless devices that operate with no connection to the building's power. They are powered via small, integral photovoltaic (solar) cells that translate the room's light into electricity. Self-powered, wireless switches are even available. These make use of the kinetic energy released when the switch is operated to charge the device's internal battery and keep it functioning.

In addition to the benefits wireless controls can bring to projects in existing buildings when they comprise a stand-alone system, these devices can also be incorporated into a wired control system on an as-needed basis when the placement of a particular device makes a physical wire run impossible. This strategy can also allow for flexibility regarding the exact placement of a control device when, as is often the case with photosensors, the devices need to be moved during commissioning once the furniture and partitions are in place and the space is completely occupied. Wireless controls can also be a good fit for a distributed system in a new construction project. In this case, all the luminaires are served with line voltage and all control system components—0-10v dimmer modules, relays,

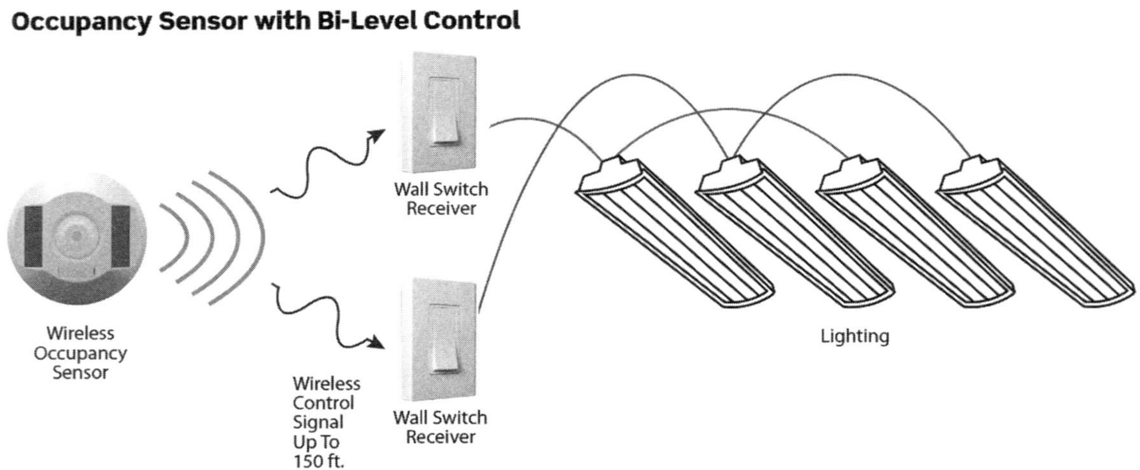


Figure 13-10. A self-powered, wireless system for simple occupancy sensor control. This system requires very little in the way of additional wiring, and is an excellent solution for a retrofit or renovation in an existing building. Image credit: Leviton Manufacturing, Inc.

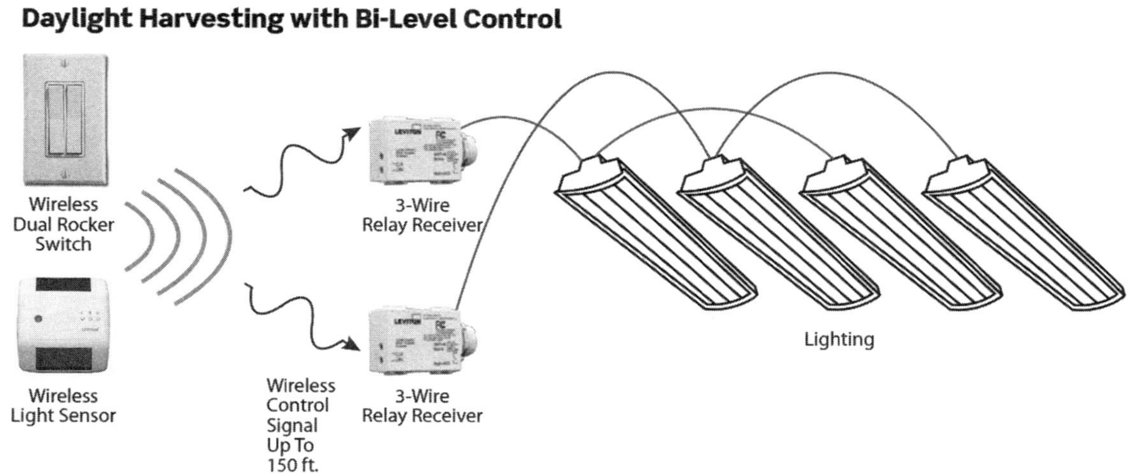


**Figure 13-11. A self-powered, wireless system for simple daylighting bi-level dimming control. This system requires very little in the way of additional wiring, and is an excellent solution for a retrofit or renovation in an existing building. Image credit: Leviton Manufacturing, Inc.**

wall-mounted controls and other kinds of input devices—exist together on a wireless network, with each one possessing a unique digital address. These are true distributed controls, as the functionality and assignment of control zones is very flexible, and entirely based on programming.

## DIMMING BALLAST AND DRIVER PERFORMANCE

Along with information on the control method, each manufacturer's ballast or driver will come with its own performance specifications. Some can reliably dim a fluorescent or LED source down to 1% of light output, and some can only dim them down to 20% before flickering, or abruptly going out. This information should be available in the product literature, and it's important to note that these numbers can be misleading. While 5 or 10% may sound quite low, our perception of light and our eyes' ability to adjust to different conditions is such that 10% of the actual light level might feel more like 30%. So whether a 1% or 20% dimming ballast in a fluorescent fixture is appropriate may have a lot to do with the application in which that fixture is being used. There is a higher expectation for the ability of lighting to dim to very low levels in an auditorium or presentation space than in a classroom or an open office, where the primary interest may be in energy savings through daylight harvesting or load-shedding.

In general, when dimming LED and fluorescent sources the best results are produced with ballasts and drivers that regulate the current to their respective sources in response to a low-voltage control signal, be it an analog 0-10 volt or any of the digital control protocols available, or via a 3-wire high-voltage signal. Fluorescent and LED luminaires with ballasts and drivers that are designed to be dimmed with standard incandescent dimmers or phase control that directly regulate the line voltage to the luminaire, will not tend to perform the best, or dim their sources to the lowest level possible. It should also be noted that in the case

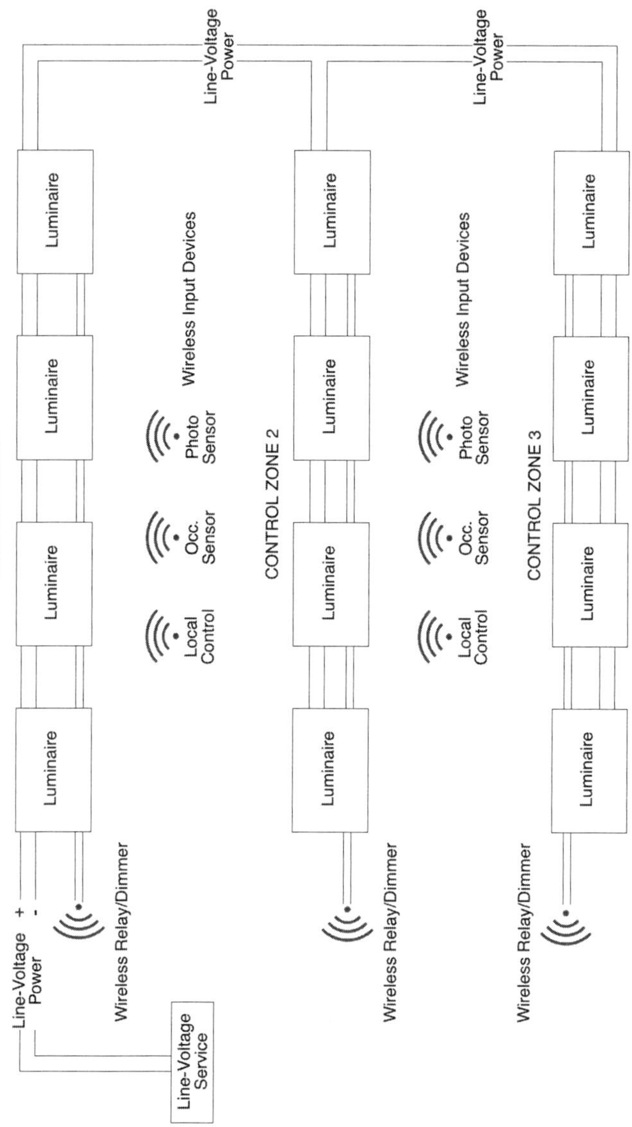


**Figure 13-12. Wiring diagram for a wireless control system.**

of fluorescent lighting different lamps dim better than others. T5's tend to be able to dim down to lower levels than T8's and compact fluorescent lamps, though this can vary from lamp to lamp and manufacturer to manufacturer. When designing a control system where dimming performance is a concern, attention must be paid to the specification of the sources, the ballasts and drivers, and the control system itself to insure the best results.

# Part V

# Building Green

# Chapter 14

# Model Codes, Code-language Standards & Energy Codes

## MODEL CODES

Model codes and code-language standards are technical documents written in a legally adoptable format that lay out the basis of a legal code, in this case a building energy code. They are developed, maintained, and published by standards organizations independently from the governing bodies with the jurisdiction and responsibility for enacting these codes. There are a number of reasons for this structure. The development of building codes can be extremely expensive, and as modern codes can be complex, more technical expertise is required to write them and to keep them up to date than is generally available to most states and municipalities. For the most part, model codes and standards, including those that mandate energy efficiency standards in our buildings, are developed and maintained by non-profit and volunteer organizations whose members contribute to the development process via a system of committees. These codes and standards are usually consensus-based, and are developed in an open, public manner, so as to insure representation from across the professional community, in a transparent process that allows input from the general public. Once a model code or standard is published, states and municipalities then have the option to adopt them as is, or with modifications, at which time they become the official code and can then be enforced by the compliance authorities within the adopting jurisdiction. The work of the model code and standard writers is generally financed by the sale of copies of the published codes and standards to municipalities, design professionals, and contractors who use it as a reference guide in the development and approval of building plans for specific projects. Model codes and standards are usually updated every three to five years, with a new version released at the time of each update. Given the time it takes for a municipality to adopt a new version of a model code or standard, it is often the case that the version currently in effect is not the latest that has been released.

The IECC (International Energy Conservation Code) is a model building energy code created by the ICC (International Code Council) that establishes minimum energy efficiency criteria for the design and construction of buildings. The ICC is a United States non-profit organization whose volunteer members come from both governmental agencies and the private sector. The IECC encourages energy conservation through efficient building envelope designs, mechanical systems, lighting systems, and the use of new materials and techniques. The IECC has different provisions for both residential and commercial buildings, and it has been adopted by many states and municipalities and integrated into their local building codes. The American Society of Heating, Refrigerating, and Air-Conditioning Engineers (ASHRAE) in conjunction with the American National Standards Institute (ANSI) and the Illuminating Engineering Society of North America (IESNA) has developed and maintains the ANSI/ASHRAE/IESNA 90.1 Energy Standard for Buildings Except Low-Rise Residential Buildings. The purpose of ANSI/ASHRAE/IESNA 90.1 is to provide minimum requirements for the energy-efficient design of the kinds of buildings it covers. It is a code-language standard that has also been adopted as a state or local energy code in many jurisdictions. The International Energy Conservation Code (IECC) also allows ASHRAE 90.1 as an alternate compliance path, making 90.1 a nearly universally accepted standard in the United States for the design of code-compliant buildings in the commercial sector. Both the IECC and ASHRAE 90.1 contain mandatory and prescriptive requirements that lay out, in a precise manner, the design parameters of various building systems, including the building envelope, HVAC, and the lighting and electrical systems. In addition to the prescriptive requirements, both of these model codes provide for a performance path to compliance that can be used in place of the prescriptive requirements. When following the performance path the building designers create an energy model (a set of calculations that predicts the energy use of a building as a whole under the proposed design), and then compare these calculations to the energy use of the same building when designed as per the minimum requirements of the prescriptive method. The performance path is a valuable tool, as it allows for more flexibility in the building design. The designers may specify individual systems that use more energy or are less efficient or are simply different than those required by the prescriptive path as long as the overall energy use of the building is demonstrated to be equal to, or less than, what it would be if the building were designed to the prescriptive specifications. On the other hand, the prescriptive path simplifies the design process by laying out, in clear terms, the steps that should be taken to assure compliance, obviating the need to do complicated calculations or create an energy model.

Currently the United States is governed by a patchwork of building energy codes based on one version or another of the IECC or ANSI/ASHRAE/IESNA 90.1. (See Figure A-7 in Appendix A for a graphic representation of the adoption of these model codes by state.) Within the states that have not adopted a statewide energy code, the local municipalities have often chosen one or the other of these codes as the basis for their local building energy code. This creates a confusing situation, so let's examine how this structure plays out in practical terms. For our example we will look at New York City, a municipality that has adopted a relatively aggressive building energy code. Starting in July 2010 New York City adopted its own energy code, the New York City Energy Conservation Code (NYCECC), which at that time (and just to confuse us a little more) was called the 2009 NYCECC. (2009 was the year it had been finalized though it was not enacted as the official code until 2010). Prior to that, New York City law required that buildings be designed and constructed to meet the statewide energy code, the Energy Conservation Construction Code of New York State (ECCCNYS). In fact, the NYCECC is essentially a modification to the 2010 ECCCNYS, which in turn relies on, and modifies, the 2009 IECC's technical provisions for the design of residential buildings and ANSI/ASHRAE/IESNA 90.1 2007's technical provisions for the design of commercial buildings. In this case, these two very influential documents, the International Energy Conservation Code and the ASHRAE 90.1 Standard, provided a baseline set of criteria, including many complex, technical requirements, that New York State and New York City could easily adopt, with some simple modifications, as their official building energy code.

## ANSI/ASHRAE/IESNA 90.1

Commonly referred to as ASHRAE 90.1, this standard provides as comprehensive a document as exists for an examination of the workings of the energy codes governing the design of lighting systems for commercial buildings in the United States. Standard 90.1 is updated by addenda that are compiled every 18 months, and is published in full every three years. Section 9 of this standard provides the minimum requirements for the energy efficiency of lighting systems. And given that the majority of states that have adopted statewide energy codes currently use the 2007 version, we will focus on the specific requirements of that version of this standard. Following is a summary of the key parts of this standard, including the prescriptive provisions, which is in no way intended as a representation of the standard in its entirety. The complete standard can be viewed and purchased on the American Society of Heating, Refrigerating, and Air-Conditioning Engineers website: www.ashrae.org.

## General Scope

The ASHRAE 90.1 Standard covers the interior spaces of buildings as well as the exterior building features, including facades, illuminated roofs, architectural features, entrances, exits, loading docks, illuminated canopies, and exterior building grounds lighting that is provided through the building's electrical service except for:

a. Emergency lighting that is automatically off during normal building operation.

b. Lighting within dwelling units.

c. Lighting that is specifically designated as required by a health or life safety statute, ordinance, or regulation.

d. Decorative gas lighting systems.

e. Alterations to existing lighting systems that replace less than 50% of the luminaires in a space provided that the alterations do not increase the installed lighting power.

## Luminaire Wattage

The installed interior lighting power must be calculated to include all power used by the luminaires, including lamps, ballasts, transformers, and control devices. (In other words, the wattage of the luminaire must be the input wattage that accounts for the losses and inefficiencies of the ballasts or drivers that drive the lamps. For luminaires that use fluorescent, HID, induction, or LED sources, the luminaire wattage can not be determined by simply adding up the rated wattage of the lamps.) Additionally, the wattage used to calculate the interior lighting power must be:

a. The maximum labeled wattage of the luminaire when different wattage lamps are supported by that luminaire.

b. For track and busway lighting that allows the addition and relocation of luminaires, the wattage must be the rated wattage of the included luminaires in the system with a minimum of 30 watts per linear foot of track or the wattage of the system's permanently wired circuit breaker or other current limiting device for line voltage systems, or the maximum wattage of the transformer supplying the system for low-voltage systems.

**Mandatory Lighting Controls**

The following lighting controls are required for a number of different space types and sizes.

a. Automatic interior lighting shutoff in buildings larger than 5000 square feet via time of day controls or occupancy sensors.

b. Independent Space Controls for break rooms, classrooms, conference rooms, and spaces of certain sizes via manual and/or occupancy sensor controls, depending on the space type and size.

c. Automatic exterior lighting shutoff controls (time of day or photosensors).

d. Master controls for hotel and motel rooms at the room entry that shut off all permanently wired lighting and switched receptacles.

e. Separate controls for task lighting, and display or demonstration lighting.

**Other Mandatory Provisions**

The following conditions must be met for these treatments and situations.

a. Tandem wiring (wiring from one luminaire to another) is required for all linear fluorescent luminaires with one or three lamps of over 30 watts each for luminaires that are ten feet apart or less (with some exceptions) in order to minimize the number of inefficient, single lamp ballasts that are employed.

b. Exit sign power is limited to 5 watts.

c. Exterior Building Grounds lighting sources over 100w must have a minimum efficacy of 60 lumens per watt or be controlled by motion sensor (with some exceptions).

**Lighting Power Densities**, or LPDs, make up the heart of these lighting energy codes. The lighting power density, usually expressed in terms of a certain number of watts per square foot ($w/ft^2$), represents the amount of power that is allowable for the lighting system per square foot of a particular space type. (Some exterior lighting, for walkways, entryways and some other treatments, will have its lighting power density defined by a certain number of watts per linear foot). By multiplying the lighting power density for a particular space or treatment by that space's overall area (or linear dimension in the case of some exterior treatments, as mentioned above) we are able to determine the lighting power allowance. So, a 1,000 square foot open office, with an allowed lighting power density of 1.1 $w/ft^2$ will have a total lighting power allowance of 1,100 watts. Different space types are allowed different lighting power densities. As an example, when using the space-by-space method for calculating interior lighting power allowances with ASHRAE 90.1, 2007, we are given 2.7 $w/ft^2$ for a hospital emergency room, and only 1.1 $w/ft^2$ for an open office. The lighting power densities are arrived at after careful study of the requirements of each space and the lighting technology that is currently available. Given this, it makes perfect sense that the emergency room would be allowed more power for lighting than the office, as there is a greater need for enhanced visual acuity when doctors and nurses are dealing with life and death medical emergencies at a fast pace. As previously mentioned, the lighting power densities must always be calculated using the luminaires' lighting input wattage that includes any ballast or driver losses.

Once the lighting power allowances are determined for all spaces and areas, they are added up to arrive at a total power allowance for the project. Depending on the specific treatment (in the case of exterior lighting), and the method used for calculating the lighting power allowances (in the case of interior lighting) there can be some tradeoffs made between spaces and areas. In this way, additional savings in one space may be used to offset an over-expenditure in another as long as the total power used by the lighting systems is less than or equal to the total power allowance for all the spaces or areas. There are two methods for calculating the interior lighting power allowances for a project: the Building Area Method, which is a simplified approach for demonstrating compliance, and the Space-by-space Method, which is an alternative approach that allows greater flexibility. Trade-offs are not allowed between spaces and areas of a building where different calculation methods are used. Following are the basic procedures to calculate a project's total lighting power allowance for each of these methods.

**Building Area Method**

a. Determine the building area type and the allowed Lighting Power Density (LPD) from the Building Area Method table provided with the standard.

b. Determine the square footage of the area.

c. Multiply the square footage of the area by that building area type's LPD to find the lighting power allowance for that area.

The interior lighting power allowance for the building is the sum of the lighting power allowances of all building area types. Trade-offs among building area types are permitted provided that the total installed interior lighting power does not exceed the interior lighting power allowance. Below is an excerpt from the lighting power density table provided for use with the building area method. [18]

| **Building Area Type** | *LPD* **(W/ft$^2$)** |
|---|---|
| Automotive facility | 0.9 |
| Convention center | 1.2 |
| Courthouse | 1.2 |
| Dining: bar lounge/leisure | 1.3 |
| Dining: cafeteria/fast food | 1.4 |
| Dining: family | 1.6 |
| Dormitory | 1.0 |

**Space-by-space Method**

a. Determine the space type and the allowed LPD from the space-by-space method table provided with the standard.

b. Determine the square footage of the space.

c. Multiply the square footage of the space by that space type's LPD to find the lighting power allowance for the space.

d. With the space–by-space method, additional lighting power allowances are available for the following treatments when they are automatically controlled, separately from the general lighting:
   1. Decorative Lighting
   2. Art & Exhibits
   3. Retail Merchandise

The interior lighting power allowance for the project is the sum of the lighting power allowances of all spaces. Trade-offs among spaces are permitted provided the total does not exceed the total interior lighting power allowance. Below is an excerpt from lighting power density table provided for use with the space-by-space method. [19]

| Common Space Types | *LPD* (W/ft$^2$) | Building-Specific Space Types | *LPD* (W/ft$^2$) |
|---|---|---|---|
| Office—Enclosed | 1.1 | <u>Gymnasium/Exercise Center</u> | |
| Office—Open Plan | 1.1 | Playing Area | 1.4 |
| Conf./Meeting/Multipurpose | 1.3 | Exercise Area | 0.9 |
| Classroom/Lecture/Training | 1.4 | <u>Courthouse/Police Station/Penitentiary</u> | |
| For Penitentiary | 1.3 | Courtroom | 1.9 |
| Lobby | 1.3 | Confinement Cells | 0.9 |
| For Hotel | 1.1 | Judges' Chambers | 1.3 |
| For Performing Arts Theater | 3.3 | <u>Fire Stations</u> | |
| For Motion Picture Theater | 1.1 | Engine Room | 0.8 |
| Audience/Seating Area | 0.9 | Sleeping Quarters | 0.3 |

**Exceptions to LPD limits:** There are some major exceptions where the lighting power associated with specific treatments is not subject to any limits, or calculated as part of a project's LPDs.

To qualify as an exception, these treatments must be separately controlled from the general lighting of the space. Examples (not a complete list) of some of these exceptions are:

- Qualifying lighting that is essential for galleries, museums, and monuments
- Lighting that is designed for and used during medical procedures
- Lighting in spaces designed for the visually impaired
- Lighting in qualifying retail display windows
- Lighting for theatrical purposes or for film and video production
- Lighting for television broadcast of sporting activities
- Lighting in casino gaming areas
- Qualifying exterior lighting in theme and amusement parks
- Furniture mounted, supplemental task lighting equipped with automatic shutoff controls

## COMCHECK & RESCHECK

COMcheck and REScheck are software applications, made available free of charge by the U.S. Department of Energy, that can be used to check the code compliance of a lighting system in commercial and residential buildings. They are available as a downloadable application that can be installed on the user's computer, or as an online, web-based application. COMcheck can be used to determine if a lighting system is code compliant in states and municipalities that rely on versions of the International Energy Conservation Code from 2000 through 2009, as well as ASHRAE 90.1 2001 through 2007. In addition, COMcheck also supports some state and local codes, which at the time of this writing include those governing Georgia, Oregon, New York, Vermont, and Pima, Arizona. REScheck can be used to determine if a lighting system is code compliant in most of the United States, with the exception of Washington, Oregon,

California Alaska, Minnesota, Indiana, and Florida. Some of the states where REScheck can be used do so by county or jurisdiction rather than statewide. For more information, go to the US Department of Energy Website: http://www.energycodes.gov/software.stm.

## CALIFORNIA'S TITLE 24

California's Title 24 is arguably the United States' most stringent energy code. It regulates residential and commercial buildings, and exceeds both the IECC and ASHRAE 90.1 in its efficiency standards for commercial buildings. And while the residential lighting provisions of the 2009 IECC simply require that 50% of the lamps in the permanently installed fixtures must be high-efficacy (defined as compact fluorescent lamps, T-8 or smaller diameter linear fluorescent lamps, or lamps with a range of minimum efficacies according to their wattage), California's Title 24 goes much farther. Following are some of the more notable additional requirements for the design of residential lighting systems in California:

- Permanently installed kitchen lighting must be high-efficacy with no standard, screw based lamps. Instead, a GU24 based lamp must be used. (This is to prohibit the owner from purchasing screw-based lamp replacements and then substituting incandescent or halogen lamps after the home has passed inspection. See Figure 14-1.)
- Lighting in bathrooms, garages, laundry rooms, closets, and utility rooms must be high efficacy; or it must be controlled by vacancy sensors in spaces that are 70 square feet or larger.
- Lighting in rooms other than bathrooms, garages, laundry rooms, closets, and utility rooms must be high efficacy, or be controlled by dimmer switches or vacancy sensors.
- Permanently installed outdoor lighting (including lighting for patios on low-rise building with four or more dwelling units, as well

as lighting for entrances, balconies, and porches) must be high efficacy or controlled by manual controls and a motion sensor and one of the following: photosensor control, astronomical time clock control, or an energy management control system.

For commercial buildings, Title 24 2008 differs from ASHRAE 90.1 2007 by requiring (partial list):

- Different methods for the calculation of interior power allowances to allow for more flexibility (Interior Area Category and Interior Tailored Methods).
- Four different lighting zones, based on population density and other factors to allow for different outdoor lighting power allowances based on different illuminance requirements in areas with greater and lesser overall exterior illumination (urban vs. rural, etc.).
- Automatic shutoff controls in all buildings with exceptions for 24 hour lighting, emergency egress, hotels/motels, high-rise residential buildings, and parking garages.
- Bi-level controls in all spaces greater than 100 square feet with lighting power densities greater than 0.8 watts/square foot.
- Separate zone of control in all daylit areas.
- Automatic daylighting controls in daylit areas greater than 2,500 square feet, with skylights (with some exceptions).

Other differences between California's Title 24 2008 and ASHRAE 90.1 2007 include:

- More categories of retail display allowances.
- Lighting power allowance credits for controls (recently added to ASHRAE 90.1 2010).
- Comprehensively addresses LED sources and luminaires.

**Figure 14-1. A self-ballasted CFL lamp with a GU24 base.**

# Chapter 15

# Paths to High Performance Buildings:

## Advocacy Groups, Advanced Energy Design Guides, Green Construction Codes & Green Building Rating Systems

### THE 2030 CHALLENGE & ZERO NET ENERGY BUILDINGS

"In the United States, the buildings sector accounted for almost 40% of primary energy consumption in 2008, 43% more than the transportation sector and 24% more than the industrial sector." [20] In 2010 the buildings sector accounted for over 40% of U.S. primary energy consumption. With this knowledge, it becomes clear to even the casual observer that the need to improve the energy-efficiency of our homes and businesses is paramount. There is no better or more effective way to reduce our carbon footprint, and our dependence on non-renewable resources, than to reduce the amount of energy required to heat, cool, and light our built environment. All efforts to reduce energy use will carry multiple benefits—environmental, economic, and political—by minimizing the negative impact of development on our natural environment, freeing us from unnecessary financial expenditures, reducing the likelihood that we will continue to pursue increasingly dangerous methods of resource extraction, and reducing the risk we will continue to become embroiled in perilous foreign policy entanglements. Add to this the potentially disastrous effects of global climate change, spurred by the combustion of the fossil fuels we most rely on—concern for which is shared by the vast majority of scientists worldwide—and the need to create a more energy-efficient building infrastructure becomes an imperative.

The **2030 Challenge**, introduced by Architecture 2030, a US based non-profit organization, and the movement for **Zero Net Energy Buildings** (ZNE), also known as Zero Energy Buildings (ZEB), both address this imperative. They are initiatives supported by a variety of advocacy groups and professional organizations, along with federal and state governmental agencies, that set out specific goals for the reduction of all new buildings' energy use and green-house gas emissions—the carbon-based gasses (mostly carbon dioxide, or $CO^2$) that are associated with global climate change. The 2030 Challenge focuses on the reduction of greenhouse gas emissions (GHGs) in new and renovated buildings by a set percentage each year so as to achieve a zero GHG standard for the design and construction of carbon neutral buildings by the year 2030. It is the hope of the proponents of this movement, which include many large international architecture and planning firms, that these target reductions will become the basis for construction and building energy codes in the near future. The movement toward Zero Net Energy buildings is related to the 2030 challenge, in that it represents an effort to reduce a building's energy use to the point where it requires no more energy than can be generated via on-site sustainable methods like solar, geothermal, and wind, all of which result in zero GHG emissions. They sound like lofty goals, but both of these initiatives are far closer to being actually possible with today's technology than is currently understood by the general public, and in fact there are zero energy buildings in existence now. Building design standards like Passivhaus (Passive House), a German standard, and MINERGIE-P, a Swiss standard, that call for super-insulated structures in conjunction with appropriate siting to make use of solar heat gain, and efficient ventilation, via heat recovery ventilators, can greatly reduce the requirements for a building's cooling and heating systems. Building to these rigorous standards and incorporating energy efficient lighting and daylighting, along with on (or off-site) generation of electricity by renewable means, can go a long way towards creating a zero energy building.

As our energy costs continue to climb, these initiatives are starting to gain traction and be reflected in our model codes and national energy programs. Numerous states have proposed or enacted legislation that mandates GHG reduction targets for publicly funded building projects, and the federal government has passed legislation requiring that new federal buildings meet the energy performance standards of the 2030 Challenge. In addition, federal legislation is pending that would require the updating of our national model building energy codes to reflect the targets of the 2030 challenge and set the United States on a path to a net-zero energy, or carbon neutral, building standard. In addition, the United States Department of Energy and the Environmental Protection Agency have sponsored a variety of programs to promote increased energy efficiency in our homes and businesses, and now provide a variety of tools and incentives to help those who want to design buildings to a higher efficiency standard than is currently mandated by law.

## ASHRAE ADVANCED ENERGY DESIGN GUIDES & ENERGY STAR

In addition creating standards for the design and construction of energy efficient buildings, the design and maintenance of ventilation systems, procedures for the commissioning of a building's mechanical systems, and many others, ASHRAE has also created a series of Advanced Energy Design Guides (AEDGs). These guides come in the form of a set of publications, by building type, that aim to help achieve a zero net energy building standard in the United States. A zero net energy building is defined by ASHRAE as "a building that, on an annual basis, draws from outside resources equal or less energy than it provides using on-site, renewable energy sources." [21] The AEDGs are a direct response to calls for a set of incremental reductions in building energy use and greenhouse gas emissions, like those proposed by the 2030 Challenge, and as such they set a progressive, escalating benchmark for the design of high performance buildings. The first set of guides, published between 2004 and 2008, had an energy savings target of 30% over ASHRAE 90.1-1999, and the second set, just beginning to become available now, have an energy savings target of 50% over ASHRAE 90.1-2004. Each guide addresses a specific building type from the following list: small office buildings, small retail buildings, K-12 schools, small warehouses and self-storage buildings, highway lodging, and small hospitals and healthcare facilities. The AEDGs provide a simplified, prescriptive approach for the design of these building types to facilitate a project team's achievement of the proposed targets. They address daylighting, simple controls, and strategies to achieve lower LPDs than those proposed by the referenced standards and mandated by the building energy code currently in effect. The Advanced Energy Design Guides are the product of a collaborative development process between a number of influential professional organizations, the U.S. government, and green building advocacy groups, including the American Institute of Architects (AIA), the Illuminating Engineering Society of North America (IESNA), the U.S. Green Building Council (USGBC), the U.S. Department of Energy (DOE), and the New Building Institute (NBI).

Energy Star is a program of the United States Environmental Protection Agency (EPA), and the Department of Energy that seeks to help individuals and businesses save money and protect the environment by promoting energy-efficient products and practices. Energy Star began as a program to rate consumer products in 1992, starting with computers and computer monitors and eventually expanding to include other office equipment, residential and commercial heating and cooling systems, and lighting. The EPA has since extended the program to cover new homes as well as commercial and industrial buildings. For the most part, the Energy Star requirements for lighting involve having Energy Star qualified fixtures as a certain percentage of the permanently installed lighting fixtures in a building. In the commercial sector, the Energy Star label signifies that a building's energy performance is better than at least 75% of similar buildings nationwide. In the residential sector, the Energy Star program for new homes provides a guide for the design and construction of homes that are at least 15% more energy-efficient than homes built to the 2004 International Residential Code, and include energy saving features that make them 20-30% more efficient than standard homes. [22]

Both the Energy Star program and the ASHRAE Advanced Energy Design Guides can provide a prescriptive compliance path for different LEED certifications from the US Green Building Council, which we will discuss in greater detail in the following chapter. Designing and constructing a commercial building as per the requirements of The Advanced Energy Design Guides may qualify a commercial new construction project for LEED credits in the Energy and Atmosphere category. And building a new home to Energy Star standards is a prescriptive path towards qualifying the home for LEED for Homes credits, also in the Energy and Atmosphere category.

## GREEN BUILDING RATING SYSTEMS

Green building rating systems are voluntary programs that owners and project teams can participate in to verify that they are designing and constructing their building projects to a high performance standard. In doing so they agree to achieve a set of sustainability benchmarks that exceed the standards of the current building codes. When a building project is registered in a particular rating system, and the team demonstrates that it has been designed and constructed to meet the appropriate benchmarks, it is granted a certification confirming that it meets that rating system's standards and is "green." These rating systems usually carry requirements that a building must meet with regard to energy use, water use, the use of renewable and sustainably produced materials for construction, siting in proximity to sensitive environments (like wetlands), design and construction techniques to reduce the impact on the surrounding environment, and limits on development in undeveloped green spaces. Additional credits are often awarded to projects that rehabilitate and reuse a previously polluted "brownfield" site so as to encourage the redevelopment of old commercial sites and minimize new development on sites that might still support agriculture, or that currently support wildlife. For lighting designers, earning credits toward a green building certification may involve reducing the load of the lighting system for a building or space so that it is lower than the allowance mandated by code, or including occupancy sensors or daylight dimming in areas where it might not otherwise be required.

But green building rating systems go a few steps further than simply "beating the code." In addition to the general requirement that buildings be designed and constructed to be as energy efficient and environmentally sustainable as possible, they also include criteria that directly address the human needs of the building's occupants. Issues like indoor environmental quality and indoor air quality are addressed. It's important to note that there is a direct relationship between indoor air quality and energy efficiency. When we start to design buildings that are tighter and tighter, with more and more effective insulation, we create an environment whereby airborne pollutants are trapped, increasing the need to use materials that do not emit toxic gasses, and to adequately ventilate the interior with fresh air. Concerns for these issues are expressed in the form of requirements that materials used in the construction of green buildings—flooring, carpeting, engineered wood products, adhesives, paints, finishes, and others—be made from non-toxic substances that do not emit pollutants, like volatile organic compounds (VOCs), into the indoor environment. For lighting, there may be a requirement that a certain percentage of a building's occupants have access to daylight and views, as this is considered a key component to an improved indoor environment. Additionally, incorporating lighting controls (bi-level switching or dimming) that allow a certain percentage of the occupants of a building or space to set the lighting levels themselves may also be encouraged, as this results in many occupants experiencing a greater degree of comfort, as well as providing a path to increased energy savings.

One of the most well-known and widely adopted programs to drive the building design and construction industry toward a green building standard is the U.S. Green Building Council's (USGBC) LEED Rating System. LEED, which is an acronym that stands for Leadership in Energy and Environmental Design, provides a tiered approach to the certification of a building as "green," with targeted versions of the rating system specifically designed for different project types including new construction, core and shell projects (for large-scale buildings), commercial interiors (for commercial tenant spaces in larger buildings), schools, retail stores, healthcare facilities, homes, neighborhood developments, and the operation and management of existing buildings. The tiered approach of the LEED rating system allows a building project to accumulate points towards certification, and depending on the number of points achieved it may earn a basic LEED certification, or a higher level LEED certification if the building has been designed and constructed to an even better standard. Points are awarded for specific components of the design and construction of the building that can be demonstrated and verified by a fairly vigorous process of documentation and submittals. If the project achieves a higher point score it may earn a LEED Silver, Gold, or Platinum certification. The structure of the LEED Rating System is organized around five main categories of green building: Sustainable Sites, Water Efficiency, Energy and Atmosphere, Materials and Resources, and Indoor Environmental Quality. The LEED rating system has been very successful in promoting of popular awareness with regard to high performance buildings. As of August 5th, 2011, there were 22,116 LEED certified projects representing a total of 1.489 billion square feet of commercial LEED certified space, which is in addition to the total area of all LEED for Homes projects. [23] Many municipalities have adopted laws requiring that certain civic buildings be designed and constructed to a specific LEED standard. But as much as LEED, and green building in general, is embraced by certain states and local jurisdictions, the national implementation of a green building standard is also hampered at times by others that still maintain outdated codes, including prohibitions on the use of grey water for sewage conveyance and on-site wastewater treatment.

The LEED rating system directly addresses lighting in a number of ways. A prerequisite for many LEED certifications in commercial buildings requires that lighting systems have a reduced connected load based on lighting power densities that are 10% below the level allowed by ASHRAE 90.1, 2007. A project's lighting design can also contribute to the points earned towards LEED certification by adhering to even lower lighting power densities that are 15-35% below that allowed by ASHRAE 90.1 (the lower the LPDs, the more points earned). Points can also be earned by including access to daylight for 75% of a building project's regularly occupied areas, access to views for 90% of a building project's regularly occupied areas, manual controls for 90% of the occupants, and daylighting controls for the first daylight zone (15′ from all windows and skylights). Additional points can be earned through the employment of daylighting controls for 50% of the connected lighting load, occupancy sensors for 75% of the connected lighting load, and for employing light-pollution reduction methods on all interior and exterior lighting.

Another, even more aggressive green building rating system is the Living Building Challenge. The Living Building Challenge, which was founded and developed by members of the Cascadia Green Building Council, is organized around seven performance areas: Site, Water, Energy, Health, Materials, Equity, and Beauty. The requirements of this rating system are challenging and steep, and include net zero energy use and net zero water use. Most of us are familiar with the technologies that can contribute to a net zero energy building, including on-site photovoltaic and solar thermal systems, hydroelectric generation, wind power, and geothermal heating and cooling. Unfortunately, the technologies that contribute to achieving, or even approaching, net zero water use are less well known, and generally face greater resistance from regulatory agencies. They include greywater systems for sewage conveyance, rainwater harvesting, and on-site wastewater treatment via the use of eco-machines. Eco-machines are biological remediation systems that employ plants and bacteria to filter wastewater in the way a wetland does, allowing the wastewater to be cleaned and re-introduced into the ground water, thus creating a closed-loop system for water use that can reduce the need for expensive and energy-hungry centralized water and sewer systems. The Living Building Challenge differs from LEED in that certification is based on actual performance, not just modeled or anticipated performance, and is only awarded after a building project has been operational for at least twelve months.

With the exception of a requirement for access to daylight for all occupiable spaces, the Living Building Challenge contains no directives regarding lighting, or that address the energy load of a project's lighting system. The requirement that all projects achieve a net zero energy status is the only related stricture, and it is mandatory. Designing lighting for a Living Building Challenge project, or any net-zero energy building, will simply require that the most aggressive energy-efficiency measures be taken at every turn, including the incorporation of high efficacy sources, high-efficiency lighting fixtures, and energy-saving controls throughout the project. Owing to its rigorous structure, and the fact that it is a relatively new rating system, The Living Building Challenge has far fewer participants to date, with three currently certified projects and more than 80 registered projects in some stage of design, construction, or operation worldwide.

## GREEN BUILDING STANDARDS & GREEN MODEL CONSTRUCTION CODES

If green building rating systems represent the leading edge in high performance building design, then green building standards and model codes are the final frontier. Green building rating systems are voluntary programs that owners and project teams can elect to follow as a means to prove that they are designing and building to a higher sustainability standard. The organizations and committees that write these systems may make use of existing standards, model codes, and design guides to create the benchmarks their rating systems are based on, but the systems are not in and of themselves written in code language, nor do they provide direct guidance as to the means and methods to achieve compliance with their stated goals. Rather, they set out minimum requirements to be met and provide for a way to confirm that the project has satisfied them. As we have discussed, some states and municipalities have enacted rules and laws mandating that certain public buildings are built to the standards laid out by LEED, and that these buildings achieve LEED certification. But to enable states and municipalities to promote the tenets of green building to the building and construction industry at large, there is a need for code-language standards and model codes that provide a clear path towards compliance with these tenets, and that can be adopted as the official building code within a particular jurisdiction. Once established, these standards can also become the baseline criteria for future rating systems, and can be adopted as the building and construction standard for private and public institutions, corporations, and other private sector organizations.

The International Green Construction Code (IGCC), currently under development by the International Code Council (ICC) is a model code whose intent is "to safeguard the environment, public health, safety and general welfare through the establishment of requirements to reduce the negative potential impacts and increase the positive potential impacts of the built environment on the natural environment and building occupants." [24] To this end the IGCC provides for minimum requirements designed to promote the conservation of natural resources, materials, and energy; the employment of renewable energy technologies; improved indoor and outdoor air quality; and sustainable practices with regards to building operations and maintenance. The IGCC is currently in the final days of the public comment phase, and should be finalized and released in early 2012.

The ASHRAE/USGBC/IESNA 189.1 Standard for the Design of High-Performance Green Buildings Except Low-Rise Residential Buildings is a code-language standard whose stated purpose is "to provide minimum requirements for the siting, design, construction, and plans for [the] operation of high performance, green buildings to: a) balance environmental responsibility, resource efficiency, occupant comfort and well being, and community sensitivity, and b) support the goal of development that meets the needs of the present without compromising the ability of future generations to meet their own needs."[25] Standard 189.1 follows a similar structure as the LEED rating system, addressing five central categories of green building:

a. site sustainability
b. water efficiency
c. energy efficiency
d. indoor environmental quality
e. building impacts on the atmosphere, materials, & resources

These five categories, in conjunction with requirements that mitigate the negative impact of construction activity, along with plans for the proper operation of a high performance building, create a "total building sustainability package for those who strive to design, build and operate green buildings." [26] As with ASHRAE 90.1 and the IECC, standard 189.1 serves as a jurisdictional compliance option to the International Green Construction Code. Lighting system requirements that must be satisfied for a project designed to this standard will follow many of the same paths we have discussed throughout this book, including reduced lighting power densities and mandatory occupancy sensing and daylighting controls in certain spaces.

# Appendix A

# Color Illustrations

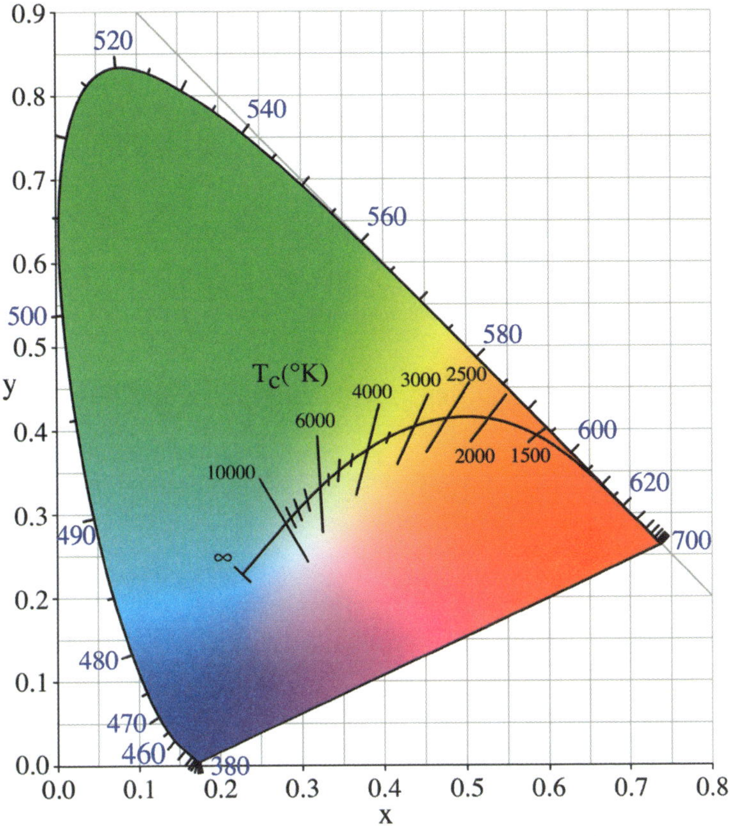


Figure A-1. The curved line in this chromaticity chart denotes the Planckian locus, which plots white light sources as they relate to the temperature of a black body radiator. A black body radiator is an ideal object that glows as it is heated, emitting light of different colors, much the way the filament in an incandescent lamp will glow when heated. As the object is first heated, it glows with a warm, amber color, which becomes more blue-white as the temperature increases. Our language for this phenomenon is counterintuitive, as we refer to the color as getting cooler, even though it relates to the light emitted by a black body radiator at a higher temperature.

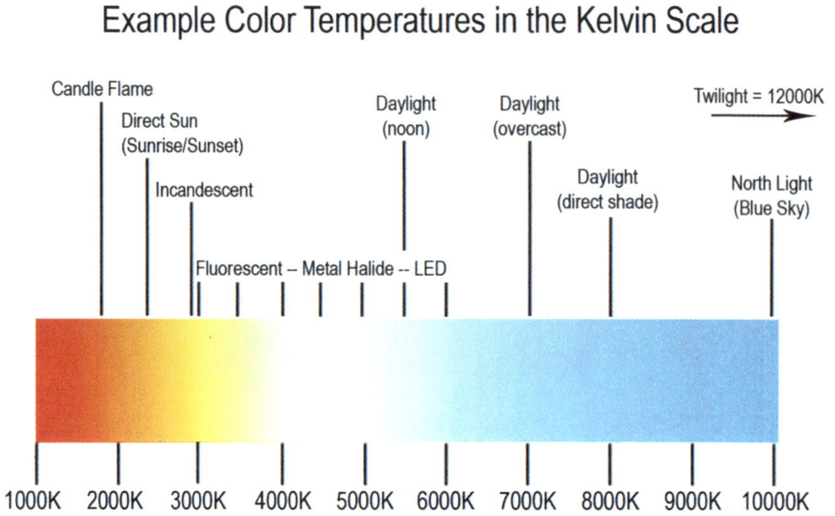


**Figure A-2. The Kelvin Scale of Correlated Color Temperature (CCT) of daylight at different times as compared to some electric light sources.**

**Figure A-3. Three lamps of different color temperatures. The self-ballasted compact fluorescent lamp on the left has a color temperature of 6500K. Both the incandescent lamp in the middle and the self-ballasted CFL on the right have color temperatures of approximately 2700K. Photo Credit: Ramjar, [CC-BY-SA 3.0 (www.creativecommons.org/licenses/by-sa/3.0)], via Wikimedia Commons**

Figure A-4. A close-up of an RGB color mixing LED lighting fixture. By combining different colors, in this case Red, Green, and Blue, the primary colors of an additive color-mixing system, this fixture can create thousands of different colors and a variety of colored lighting effects.

**Figure A-5. Arrays of small RGB LED lighting fixtures can be configured to function as low-resolution image display systems, as in the lobby installation pictured above. Photo credit: Moritz Wade/perceptual.de provided courtesy of LightWild. Architects and Designers: Populous, JSK Architects, Tropp Lighting Design, Schmidhuber & Partner, and ICON Venue Group.**

**Figure A-6. A high brightness LED creates white light by a process in which a yellow phosphor is excited by a colored LED, often a royal blue.**

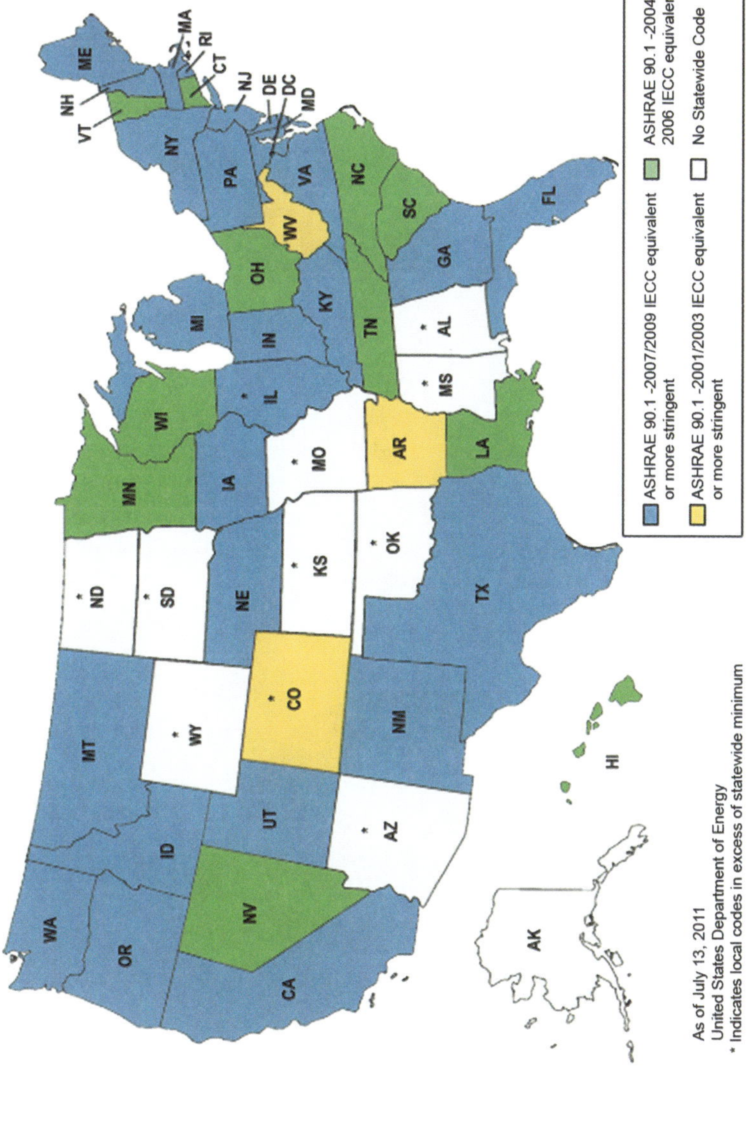


Figure A-7. Adoption of model energy codes by state.

# Appendix B

# Lighting Calculations and Calculation Software

**Simplified point-to-point calculation based on the Inverse Square Law* (does not account for the inter-reflection of light within a space).**

The calculation of footcandles at a point, no matter if it is on a horizontal, vertical or tilted surface, can be accomplished with the inverse square law. The law states that the illuminance is proportional to the candlepower of the source in the given direction and inversely proportional to the square of the distance from the source. In addition, as a surface is tilted away from the source, illuminance will decrease in a ratio equal to the cosine of the angle of incidence. The inverse square law formula can be expressed in various ways, and the two most useful follow:

> Version 1 (Figure B-1) is ideal for the complexities of three-dimensional space—no trigonometry (cosine) is needed, just the simple X, Y, and Z coordinates of the layout. It is also very useful in calculating foot-candles from the CBCP (center beam candlepower) of accent lights.
>
> Version 2 (Figure B-2) is useful for calculations that can be laid out in two dimensions, and when it is easy to find the cosine of the aiming angle.

Insert your data into either formula to calculate the initial footcandle level at a point. (FC = Footcandles, CD = Candela)

**Version 1:** $FC = \dfrac{I \times N}{D^3}$

Where:

I = cd towards point

D = distance to point (the square root of $X^2 + Y^2 + Z^2$)

N = distance normal to the surface

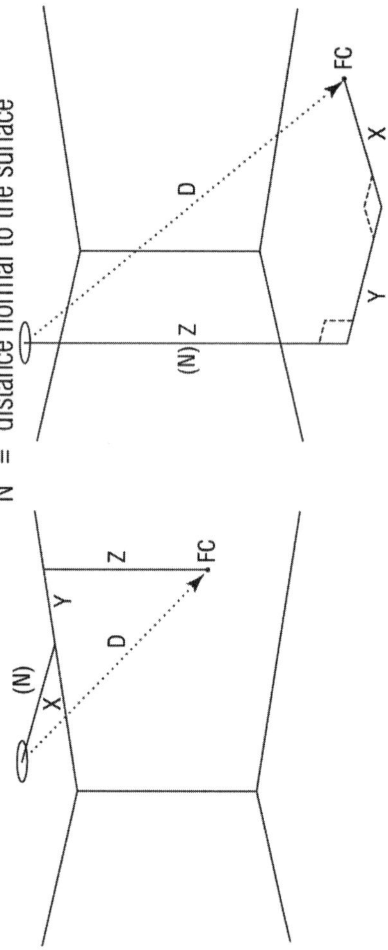


**Figure B-1. Point by Point Calculation Method Version 1**

**Version 2:** $FC = \dfrac{I \times \cos\theta}{D^2}$

Where:

$I$ = cd towards point

$D$ = distance to point (the square root of $X^2 + Y^2$)

$\theta$ = angle between incident light ray and normal to the surface

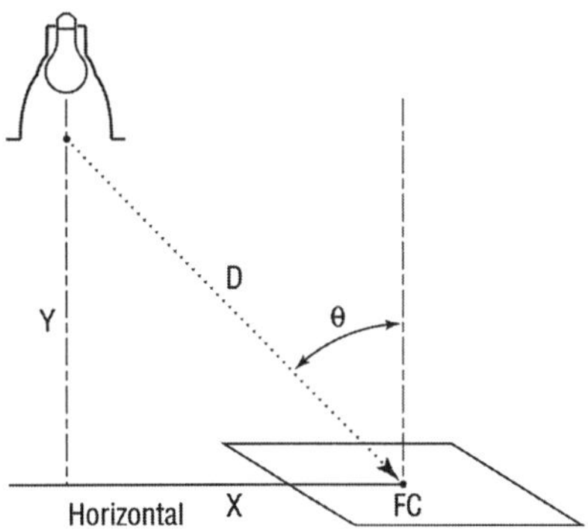


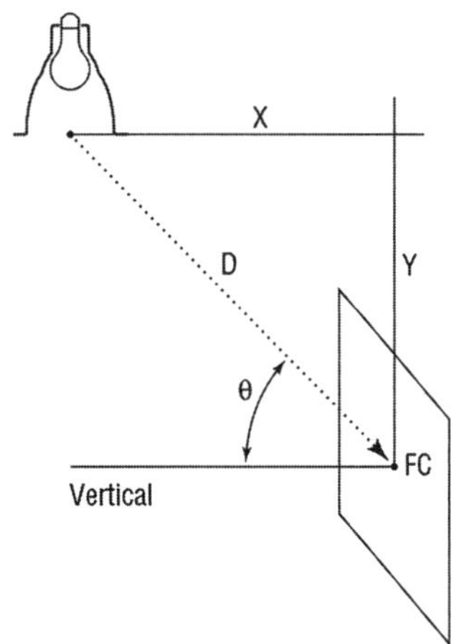


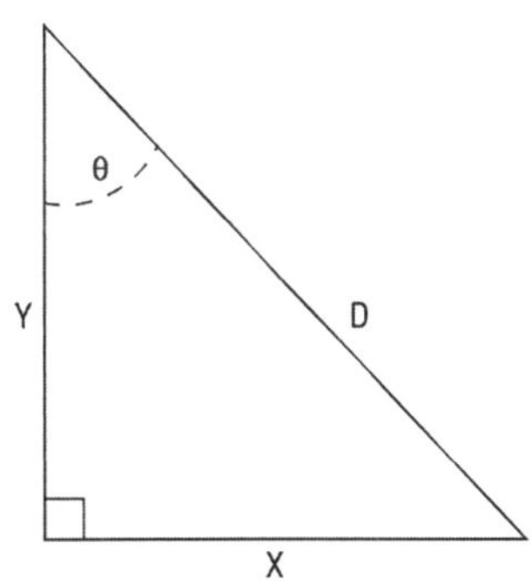


| θ | cos θ | TAN θ |
|---|---|---|
| 0° | 1.0 | 0.0 |
| 5° | .996 | .087 |
| 10° | .985 | .176 |
| 15° | .966 | .268 |
| 20° | .940 | .364 |
| 25° | .906 | .466 |
| 30° | .866 | .577 |
| 35° | .819 | .700 |
| 40° | .766 | .839 |
| 45° | .707 | 1.0 |

To find distance D: $D = \sqrt{X^2 + Y^2}$ or $D = \dfrac{Y}{\cos\theta}$

To find aiming angle θ: $\tan\theta = \dfrac{X}{Y}$ or $\cos\theta = \dfrac{Y}{D}$

**Figure B-2. Point by Point Calculation Method Version 2**
***Point-to-point Calculations & Diagrams Courtesy of Cooper Lighting**

**Zonal cavity method (for regular layouts of luminaires in regularly shaped rooms, considers the inter-reflection of light between room surfaces)**

The zonal cavity method (sometimes called the Lumen Method) is an accepted method for calculating average illuminance levels for regularly shaped indoor spaces, unless the light distribution is radically asymmetric. It is an accurate hand method because it takes into consideration the effect that inter-reflectance has on the overall average illuminance. The basis of the zonal cavity method is that a room is made up of three spaces, or cavities. The space between the ceiling and the fixtures, if they are suspended, is defined as the *ceiling cavity*, the space between the work-plane and the floor is the *floor cavity*, and the space between the fixtures and the work-plane is the *room cavity*. Once these cavities are defined, numerical relationships, called *cavity ratios*, can be calculated and used to determine the *effective reflectance* of the ceiling and floor cavities. With this information we can find the *coefficient of utilization* for the luminaires to be installed, from a chart published by the luminaire manufacturer, and then compute the average illuminance in the space. There are four basic steps:

1. Determine the cavity ratios
2. Determine effective cavity reflectances
3. Select coefficient of utilization
4. Compute average illuminance level

**Step 1 (see Figure B-3): Calculate the cavity ratios for a rectangular space by using the following formulas:**

$$\text{Ceiling Cavity Ratio (CCR)} = \frac{5\ \text{hcc}\ (L+W)}{L \times W}$$

$$\text{Room Cavity Ratio (RCR)} = \frac{5\ \text{hrc}\ (L+W)}{L \times W}$$

$$\text{Floor Cavity Ratio (FCR)} = \frac{5\ \text{hfc}\ (L+W)}{L \times W}$$

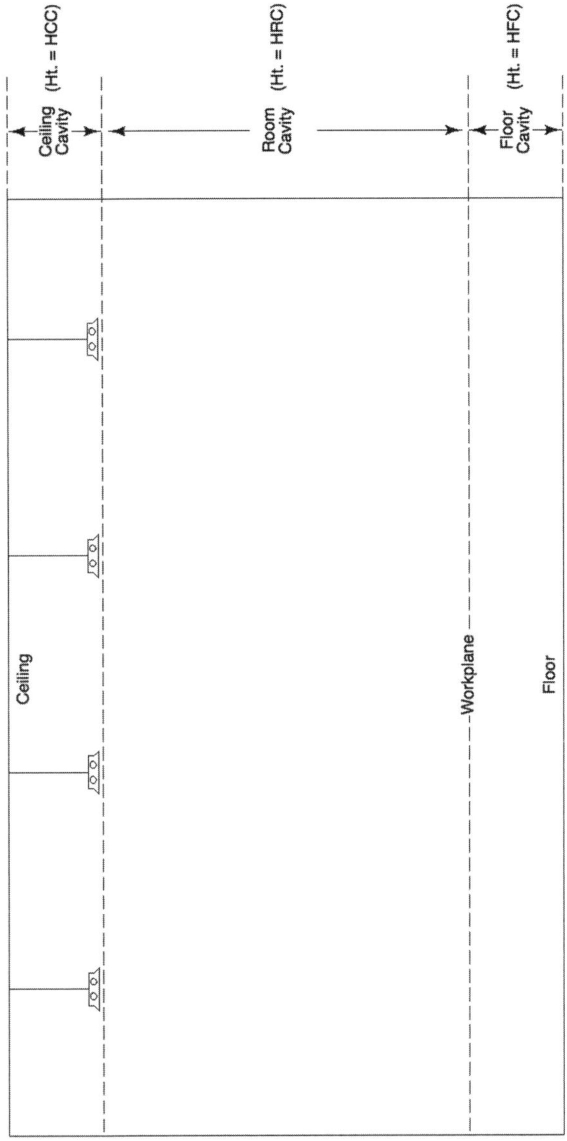


**Figure B-3. Calculating cavity ratios for a rectangular space**

**Table A: Percent effective ceiling or floor cavity reflectance for various reflectance combinations.**

| % Ceiling or Floor Reflectance | 90 | | | | 80 | | | | 70 | | | 50 | | | | 30 | | | 10 | | |
|---|---|---|---|---|---|---|---|---|---|---|---|---|---|---|---|---|---|---|---|---|---|
| % Wall Reflectance | 90 | 70 | 50 | 30 | 80 | 70 | 50 | 30 | 70 | 50 | 30 | 70 | 50 | 30 | 70 | 50 | 30 | 10 | 50 | 30 | 10 |
| **Cavity ratio** | | | | | | | | | | | | | | | | | | | | | |
| 0.2 | 89 | 88 | 86 | 85 | 78 | 78 | 77 | 76 | 68 | 67 | 66 | 49 | 48 | 47 | 30 | 29 | 29 | 28 | 10 | 10 | 09 |
| 0.4 | 88 | 86 | 84 | 81 | 77 | 76 | 74 | 72 | 67 | 65 | 63 | 48 | 47 | 45 | 30 | 29 | 28 | 26 | 11 | 10 | 09 |
| 0.6 | 87 | 84 | 80 | 77 | 76 | 75 | 71 | 68 | 65 | 63 | 59 | 47 | 45 | 43 | 30 | 28 | 26 | 25 | 11 | 10 | 08 |
| 0.8 | 87 | 82 | 77 | 73 | 75 | 73 | 69 | 65 | 64 | 60 | 56 | 47 | 44 | 40 | 30 | 28 | 25 | 23 | 11 | 10 | 08 |
| 1.0 | 86 | 80 | 75 | 69 | 74 | 72 | 67 | 62 | 62 | 58 | 53 | 46 | 43 | 38 | 30 | 27 | 24 | 22 | 12 | 10 | 08 |
| 1.2 | 85 | 78 | 72 | 66 | 73 | 70 | 64 | 58 | 61 | 57 | 50 | 45 | 41 | 36 | 30 | 27 | 23 | 21 | 12 | 10 | 07 |
| 1.4 | 85 | 77 | 69 | 62 | 72 | 68 | 62 | 55 | 60 | 55 | 47 | 45 | 40 | 35 | 30 | 26 | 22 | 19 | 12 | 10 | 07 |
| 1.6 | 84 | 75 | 67 | 59 | 71 | 67 | 60 | 53 | 59 | 53 | 45 | 44 | 39 | 33 | 29 | 25 | 22 | 18 | 12 | 09 | 07 |
| 1.8 | 83 | 73 | 64 | 56 | 70 | 66 | 58 | 50 | 58 | 51 | 42 | 43 | 38 | 31 | 29 | 25 | 21 | 17 | 13 | 09 | 06 |
| 2.0 | 83 | 72 | 62 | 53 | 69 | 64 | 56 | 48 | 56 | 49 | 40 | 43 | 37 | 30 | 29 | 24 | 20 | 16 | 13 | 09 | 06 |
| 2.2 | 82 | 70 | 59 | 50 | 68 | 63 | 54 | 45 | 55 | 48 | 38 | 42 | 36 | 29 | 29 | 24 | 19 | 15 | 13 | 09 | 06 |
| 2.4 | 82 | 69 | 58 | 48 | 67 | 61 | 52 | 43 | 54 | 46 | 37 | 42 | 35 | 27 | 29 | 24 | 19 | 14 | 13 | 09 | 06 |
| 2.6 | 81 | 67 | 56 | 46 | 66 | 60 | 50 | 41 | 54 | 45 | 35 | 41 | 34 | 26 | 29 | 23 | 18 | 14 | 13 | 09 | 06 |
| 2.8 | 81 | 66 | 54 | 44 | 65 | 59 | 48 | 39 | 53 | 43 | 33 | 41 | 33 | 25 | 29 | 23 | 17 | 13 | 13 | 09 | 05 |
| 3.0 | 80 | 64 | 52 | 42 | 65 | 58 | 47 | 37 | 52 | 42 | 32 | 40 | 32 | 24 | 29 | 22 | 17 | 12 | 13 | 09 | 05 |
| 3.2 | 79 | 63 | 50 | 40 | 65 | 57 | 45 | 35 | 51 | 40 | 31 | 39 | 31 | 23 | 29 | 22 | 16 | 12 | 13 | 09 | 05 |
| 3.4 | 79 | 62 | 48 | 38 | 64 | 56 | 44 | 34 | 50 | 39 | 29 | 39 | 30 | 22 | 29 | 22 | 16 | 11 | 13 | 09 | 05 |
| 3.6 | 78 | 61 | 47 | 36 | 63 | 54 | 43 | 32 | 49 | 38 | 28 | 39 | 29 | 21 | 29 | 21 | 15 | 10 | 13 | 09 | 04 |
| 3.8 | 78 | 60 | 45 | 35 | 62 | 53 | 41 | 31 | 49 | 37 | 27 | 38 | 29 | 21 | 28 | 21 | 15 | 10 | 14 | 09 | 04 |
| 4.0 | 77 | 58 | 44 | 33 | 61 | 53 | 40 | 30 | 48 | 36 | 26 | 38 | 28 | 20 | 28 | 21 | 14 | 09 | 14 | 09 | 04 |
| 4.2 | 77 | 57 | 43 | 32 | 60 | 52 | 39 | 29 | 47 | 35 | 25 | 37 | 28 | 20 | 28 | 20 | 14 | 09 | 14 | 09 | 04 |
| 4.4 | 76 | 56 | 42 | 31 | 60 | 51 | 38 | 28 | 46 | 34 | 24 | 37 | 27 | 19 | 28 | 20 | 14 | 09 | 14 | 08 | 04 |
| 4.6 | 76 | 55 | 40 | 30 | 59 | 50 | 37 | 27 | 45 | 33 | 24 | 36 | 26 | 18 | 28 | 20 | 13 | 08 | 14 | 08 | 04 |
| 4.8 | 75 | 54 | 39 | 28 | 58 | 49 | 36 | 26 | 45 | 32 | 23 | 36 | 26 | 18 | 28 | 20 | 13 | 08 | 14 | 08 | 04 |
| 5.0 | 75 | 53 | 38 | 28 | 58 | 48 | 35 | 25 | 44 | 31 | 22 | 35 | 25 | 17 | 28 | 19 | 13 | 08 | 14 | 08 | 04 |

Figure B-4. Zonal Cavity Calculation Table A

Where:

hcc = distance in feet from luminaire to ceiling
hrc = distance in feet from luminaire to work-plane
hfc = distance in feet from work-plane to floor
L = length of room, in feet
W = width of room, in feet

An alternate formula for calculating any cavity ratio is:

$$\text{Cavity Ratio} = \frac{2.5 \text{ x height of cavity x cavity perimeter}}{\text{area of cavity base}}$$

**Step 2: Determine effective cavity reflectances for the ceiling and floor cavities.**

These can be found in Table A (Figure B-4) under the applicable combination of cavity ratio and actual reflectance of ceiling, walls, and floor. The effective reflectance values found will then be ρcc (effective ceiling cavity reflectance) and ρfc (effective floor cavity reflectance). Note that if the luminaire is recessed or surface mounted, or if the floor is the work-plane, the ceiling cavity ratio (CCR) or floor cavity ratio (FCR) will be 0 and the actual reflectance of the ceiling or floor will be the effective reflectance.

**Step 3:** With the values for ρcc, ρfc, and ρw* (wall reflectance), and the room cavity ratio (RCR) previously calculated, find the coefficient of utilization for the luminaire to be installed in the coefficient of utilization (CU) table published by the luminaire manufacturer. An example of a CU table can be found in Figure B-5.

The coefficient of utilization found will be for a 20% effective floor cavity reflectance. Thus, it will be necessary to correct for the previously determined ρfc if it is not 20 by multiplying the previously determined CU by the appropriate factor from Table B (Figure B-6).

---

*NB: Sometimes the Greek letter ρ is used to represent reflectance, and sometimes a lowercase p, because ρ looks like p. And sometimes R is used, as the first letter in reflectance (see Chapter 6).

**COEFFICIENT OF UTILIZATION TABLE**

Effective Floor cavity Reflectance = 20%

| Pcc ... | 80 | | | | 70 | | | | 50 | | | 30 | | | 10 | | | 0 |
|---|---|---|---|---|---|---|---|---|---|---|---|---|---|---|---|---|---|---|
| Pw ... | 70 | 50 | 30 | 10 | 70 | 50 | 30 | 10 | 50 | 30 | 10 | 50 | 30 | 10 | 50 | 30 | 10 | 0 |
| RCR | | | | | | | | | | | | | | | | | | |
| 0 | 1.02 | 1.02 | 1.02 | 1.02 | 1.00 | 1.00 | 1.00 | 1.00 | .95 | .95 | .95 | .91 | .91 | .91 | .87 | .87 | .87 | .86 |
| 1 | .94 | .90 | .86 | .83 | .91 | .88 | .85 | .82 | .84 | .81 | .79 | .81 | .79 | .77 | .78 | .76 | .74 | .73 |
| 2 | .85 | .78 | .73 | .68 | .83 | .77 | .72 | .67 | .74 | .69 | .66 | .71 | .67 | .64 | .68 | .65 | .63 | .61 |
| 3 | .78 | .69 | .62 | .57 | .76 | .68 | .61 | .56 | .65 | .60 | .55 | .63 | .58 | .54 | .60 | .57 | .53 | .51 |
| 4 | .71 | .61 | .54 | .48 | .69 | .60 | .53 | .48 | .58 | .52 | .47 | .56 | .51 | .46 | .54 | .49 | .46 | .44 |
| 5 | .65 | .55 | .47 | .41 | .64 | .54 | .46 | .41 | .52 | .45 | .41 | .50 | .45 | .40 | .48 | .44 | .40 | .38 |
| 6 | .60 | .49 | .41 | .36 | .59 | .48 | .41 | .36 | .47 | .40 | .36 | .45 | .40 | .35 | .44 | .39 | .35 | .33 |
| 7 | .56 | .45 | .37 | .32 | .55 | .44 | .37 | .32 | .42 | .36 | .31 | .41 | .36 | .31 | .40 | .35 | .31 | .29 |
| 8 | .52 | .41 | .33 | .28 | .51 | .40 | .33 | .28 | .39 | .33 | .28 | .38 | .32 | .28 | .37 | .32 | .28 | .26 |
| 9 | .49 | .37 | .30 | .26 | .47 | .37 | .30 | .25 | .36 | .30 | .25 | .35 | .29 | .25 | .34 | .29 | .25 | .23 |
| 10 | .46 | .34 | .28 | .23 | .45 | .34 | .27 | .23 | .33 | .27 | .23 | .32 | .27 | .23 | .32 | .26 | .23 | .21 |

**Figure B-5. Coefficient of Utilization Table for a Representative Luminaire**

**Table B: Multiplying factors for other than 20 percent effective floor cavity reflectance**

| % Effective Ceiling Cavity Reflectance (pcc) | 80 | | | | 70 | | | | 50 | | | 30 | | | 10 | | |
|---|---|---|---|---|---|---|---|---|---|---|---|---|---|---|---|---|---|
| % Wall Reflectance (pw) | 70 | 50 | 30 | 10 | 70 | 50 | 30 | 10 | 50 | 30 | 10 | 50 | 30 | 10 | 50 | 30 | 10 |

For 30 per cent effective floor cavity reflectance (20 per cent = 1.00)

| Room Cavity Ratio | | | | | | | | | | | | | | | | | |
|---|---|---|---|---|---|---|---|---|---|---|---|---|---|---|---|---|---|
| 1 | 1.092 | 1.082 | 1.075 | 1.068 | 1.077 | 1.070 | 1.064 | 1.059 | 1.049 | 1.044 | 1.040 | 1.028 | 1.026 | 1.023 | 1.012 | 1.010 | 1.008 |
| 2 | 1.079 | 1.066 | 1.055 | 1.047 | 1.068 | 1.057 | 1.048 | 1.039 | 1.041 | 1.033 | 1.027 | 1.026 | 1.021 | 1.017 | 1.013 | 1.010 | 1.006 |
| 3 | 1.070 | 1.054 | 1.042 | 1.033 | 1.061 | 1.048 | 1.037 | 1.028 | 1.034 | 1.027 | 1.020 | 1.024 | 1.017 | 1.012 | 1.014 | 1.009 | 1.005 |
| 4 | 1.062 | 1.045 | 1.033 | 1.024 | 1.055 | 1.040 | 1.029 | 1.021 | 1.030 | 1.022 | 1.015 | 1.022 | 1.015 | 1.010 | 1.014 | 1.009 | 1.004 |
| 5 | 1.056 | 1.038 | 1.026 | 1.018 | 1.050 | 1.034 | 1.024 | 1.015 | 1.027 | 1.018 | 1.012 | 1.020 | 1.013 | 1.008 | 1.014 | 1.009 | 1.004 |
| 6 | 1.052 | 1.033 | 1.021 | 1.014 | 1.047 | 1.030 | 1.020 | 1.012 | 1.024 | 1.015 | 1.009 | 1.019 | 1.012 | 1.006 | 1.014 | 1.008 | 1.003 |
| 7 | 1.047 | 1.029 | 1.018 | 1.011 | 1.043 | 1.026 | 1.017 | 1.009 | 1.022 | 1.013 | 1.007 | 1.018 | 1.010 | 1.005 | 1.014 | 1.008 | 1.003 |
| 8 | 1.044 | 1.026 | 1.015 | 1.009 | 1.040 | 1.024 | 1.015 | 1.007 | 1.020 | 1.012 | 1.006 | 1.017 | 1.009 | 1.004 | 1.013 | 1.007 | 1.003 |
| 9 | 1.040 | 1.024 | 1.014 | 1.007 | 1.037 | 1.022 | 1.014 | 1.006 | 1.019 | 1.011 | 1.005 | 1.016 | 1.009 | 1.004 | 1.013 | 1.007 | 1.002 |
| 10 | 1.037 | 1.022 | 1.012 | 1.006 | 1.034 | 1.020 | 1.012 | 1.005 | 1.017 | 1.010 | 1.004 | 1.015 | 1.009 | 1.003 | 1.013 | 1.007 | 1.002 |

For 10 per cent effective floor cavity reflectance (20 per cent=1.00)

| Room Cavity Ratio | | | | | | | | | | | | | | | | | |
|---|---|---|---|---|---|---|---|---|---|---|---|---|---|---|---|---|---|
| 1 | .923 | .929 | .935 | .940 | .933 | .939 | .943 | .948 | .956 | .960 | .963 | .973 | .976 | .979 | .989 | .991 | .993 |
| 2 | .931 | .942 | .950 | .958 | .940 | .949 | .957 | .963 | .962 | .968 | .974 | .976 | .980 | .985 | .988 | .991 | .995 |
| 3 | .939 | .951 | .961 | .969 | .945 | .957 | .966 | .973 | .967 | .975 | .981 | .978 | .983 | .988 | .988 | .992 | .996 |
| 4 | .944 | .958 | .969 | .978 | .950 | .963 | .973 | .980 | .972 | .980 | .986 | .980 | .986 | .991 | .987 | .992 | .996 |
| 5 | .949 | .964 | .976 | .983 | .954 | .968 | .978 | .985 | .975 | .983 | .989 | .981 | .988 | .993 | .987 | .992 | .997 |
| 6 | .953 | .969 | .980 | .986 | .958 | .972 | .982 | .989 | .977 | .985 | .992 | .982 | .989 | .995 | .987 | .993 | .997 |
| 7 | .957 | .973 | .983 | .991 | .961 | .975 | .985 | .991 | .979 | .987 | .994 | .983 | .990 | .996 | .987 | .993 | .998 |
| 8 | .960 | .976 | .986 | .993 | .963 | .977 | .987 | .993 | .981 | .988 | .995 | .984 | .991 | .997 | .987 | .994 | .998 |
| 9 | .963 | .978 | .987 | .994 | .965 | .979 | .989 | .994 | .983 | .990 | .996 | .985 | .992 | .998 | .988 | .994 | .999 |
| 10 | .965 | .980 | .965 | .980 | .967 | .981 | .990 | .995 | .984 | .991 | .997 | .986 | .993 | .998 | .988 | .994 | .999 |

**Figure B-6. Zonal Cavity Calculation Table B (factors for other than 20% effective floor cavity reflectance)**

The final CU is the CU (20% floor) x the multiplier for the actual ρfc. If the ρfc is other than 10% or 30%, it will be necessary to interpolate or extrapolate the appropriate multiplier.

**Step 4: Compute the illuminance level using the standard Lumen Method formula.**

Footcandles (maintained) =

$$\frac{(\#\text{ of fixtures x lamps per fixture) x (lumens per lamp) x CU x LLF}}{\text{area in square feet}}$$

Where:

LLF = Light Loss Factor

When the initial illuminance level required is known, and the intention is to compute the number of fixtures needed to obtain that level, a variation of the standard lumen formula may be used.

# of luminaires =

$$\frac{\text{maintained footcandles desired x area in square feet.}}{\text{(lamps per fixture) x (lumens per lamp) x CU x LLF}}$$

The total light loss factor (LLF) consists of three basic factors: lamp lumen depreciation (LLD), luminaire dirt depreciation (LDD), and ballast factor (BF). To calculate initial levels, a multiplier of 1 is used. Light loss factors, along with the maintained or mean lamp lumen output, vary with manufacturer and type of lamp or luminaire, and are determined by consulting the manufacturer's published data.

Ballast factor (BF) is defined as the ratio between the published lamp lumens and the lumens delivered by the lamp on the ballast used. Occasionally, other light loss factors may need to be applied. Some of these are luminaire ambient temperature, voltage factor, and room surface dirt depreciation.

**Commonly Available Lighting Calculation & 3D Rendering Computer Software**

**AGi32** (Lighting Analysts)
http://www.agi32.com

**Elum Tools** (Lighting Analysts)
http://www.elumtools.com.

**DIALux** (DIAL)
http://www.dial.de/CMS/English/Articles/DIALux/Download/Download_d_e_fr_it_es_cn.html#

**Radiance** (Lawrence Berkeley National Laboratory)
http://radsite.lbl.gov/radiance/HOME.html

**3D Studio Max** (Autodesk)
http://usa.autodesk.com/3ds-max/

**Vectorworks** (Nemetschek)
http://www.nemetschek.net/

**Luminaire Manufacturers Providing Online Indoor Zonal Cavity Calculations and/or Outdoor Point-to-Point Calculations via Flash_Indoor or Flash_Outdoor (may be a partial list):**

Artemide

Color Kinetics (Philips)
Cooper Lighting
Crescent/Stonco (Philips)
Day-Brite (Philips)
Day-Brite Canada
EGS Electrical Group (Appleton)
Exceline (Philips)
HE Williams
HML Limited (New Zealand)
Hubbell
Juno Lighting Group
Kurt Versen
Lightolier (Philips)
Nicor Lighting
Orion Energy
RAB Lighting
Simkar
Siteco GmbH
Tamlite

# Appendix C

# Resources

Illumination Engineering Society of North America (IESNA), http://www.iesna.org/

Association of Energy Engineers (AEE), http://www.aeecenter.org/

Consortium for Energy Efficiency (CEE), http://www.cee1.org/

American Society of Heating, Refrigerating and Air-Conditioning Engineers (ASHRAE), http://www.ashrae.org/

U.S. Green Building Council (USGBC), http://www.usgbc.org/

Living Building Challenge, https://ilbi.org/lbc

Architecture 2030, http://www.architecture2030.org/

DOE CALiPER Program: testing results for a wide array of SSL products available for general illumination. http://www1.eere.energy.gov/buildings/ssl/caliper.html

Next Generation Luminaires Design Competition: Solid State Lighting (SSL) Design Competition created to recognize and promote excellence in the design of energy-efficient LED commercial lighting luminaires. http://www.ngldc.org/

Energy Star: Joint program of the U.S. Environmental Protection Agency and the U.S. Department of Energy with programs and ratings designed to protect the environment and save money through energy efficient products and practices. http://www.energystar.gov/

REFERENCED CODES, STANDARDS, & GUIDES:

ANSI/ASHRAE/IESNA 90.1, Energy Standard for Buildings Except Low-Rise Residential Buildings (ANSI Approved; IESNA Co-Sponsored)

ASHRAE/USGBC/ISNA 189.1, Standard for the Design of High-Performance Green Buildings (ANSI Approved; USGBC and IES Co-sponsored)

IECC (International Energy Conservation Code): Model energy building code with minimum energy efficiency provisions for residential and commercial buildings, offering both prescriptive- and performance-based approaches.

IGCC (International Green Construction Code): Model construction code with minimum regulations for buildings and systems, offering both prescriptive- and performance-based approaches, to safeguard the environment, public health, safety and general welfare through the establishment of requirements to reduce the negative potential impacts and increase the positive potential impacts of the built environment on the natural environment and building occupants, by means of minimum requirements related to: conservation of natural resources, materials and energy; the employment of renewable energy technologies, indoor and outdoor air quality; and building operations and maintenance.

Title 24: California's Energy Efficiency Standards for Residential and Nonresidential Buildings

ASHRAE Advanced Energy Design Guides: recommendations for achieving energy savings over the minimum code requirements of ANSI/ ASHRAE/IESNA Standard 90.1

## REFERENCES


1. Illuminating Engineering Society of North America (IESNA), American National Standard Practice for Office Lighting, ANSI/ IESNA-RP-1-04, New York, NY, 2004.
2. DiLaura, Houser, Mistrick & Steffy, The Illuminating Engineering Society Lighting Handbook, 10th Edition, 2011, Pg. 12.17
3. *Ibid*, Pg. 12.20
4. Davis & Ohno, Development of a Color Quality Scale, National Institute of Standards & Technology, 2006
5. Boyce, Veitch, Newsham, Myer, & Hunter, Lighting Quality and Office Work, US Dept. of Energy, 2003, Pg. 162
6. Mills & Borg, Trends In Recommended Illuminance Levels: An International Comparison, (Journal of the Illuminating Engineering Society, Winter 1999)
7. P.R. Boyce, Human Factors In Lighting, (Taylor & Francis, 2003), pp 227
8. DiLaura, Houser, Mistrick & Steffy, The Illuminating Engineering Society Lighting Handbook, 10th Edition, 2011, Pg. 32.4-32.6
9. *Ibid*, Pg 4.30
10. *Ibid*, Chapters 21-37

11. Roberts, How Magnetic Induction Lamps Work, InduLux Technologies, 2011
12. Lighting Research Center, Rensselaer Polytechnic Institute, Capturing the Daylight Dividend, 2006
13. Jenkins, Phillips, Mulberg, Hui, Activity patterns of Californians: Use of and proximity to indoor pollutant sources, Atmospheric Environment. Part A. General Topics, Volume 26, Issue 12, August 1992, Pages 2141-2148
14. Shin, Yun, Kim, Influences of Subjective Assessments of Discomfort Glare from Windows on Lighting Energy Use, Department of Architectural Engineering, Kyung Hee University
15. New York State Energy Research & Development Authority, Commercial Lighting Program Advanced Lighting Training, 2010
16. Lighting Research Center, Rensselaer Polytechnic Institute, Reducing Barriers to Use of High Efficiency Lighting Systems, 2004
17. Boyce, Veitch, Newsham, Myer, & Hunter, *op cit*
18. ©2007, ASHRAE (www.ashrae.org). Used with permission from ASHRAE Standard/IESNA 90.1-2007: Energy Standard for Buildings Except Low-Rise Residential Buildings IP Version, Section 9, Lighting
19. *Ibid*
20. United States Department of Energy, Buildings Energy Data Book, 2010: http://buildingsdatabook.eren.doe.gov/ChapterIntro1.aspx
21. American Society of Heating, Refrigerating, and Air-Conditioning Engineers, Advanced Energy Design Guides
22. Energy Star Website, (New Homes & Buildings and Plants), 2011: http://www.energystar.gov/index.cfm?c=new_homes.hm_index http://www.energystar.gov/index.cfm?c=business.bus_index
23. U.S. Green Building Council Website, August 5th, 2011
24. International Code Council, International Green Construction Code, Public Version 2.0, 2010
25. American Society of Heating, Refrigerating, and Air-Conditioning Engineers, Standard 189.1 For the Design of High-Performance Green Buildings Except Low-Rise Residential Buildings, Purpose
26. American Society of Heating, Refrigerating, and Air-Conditioning Engineers Website, The Green Standard: http://www.ashrae.org/publications/page/927

# Index

**Symbols**
0-10v 174, 175
2030 Challenge 204, 206

**A**
absolute photometry 107, 120, 122
Advanced Energy Design Guides 205, 206, 207
American Society of Heating, Refrigerating, and Air-Conditioning Engineers 192, 194
ASHRAE 192
ASHRAE 90.1 192, 194, 195, 201
ASHRAE/USGBC/IESNA 189.1 213
astronomical clocks 164

**B**
ballast 111
ballast efficiency factor 99
ballast factor 97
binning 117
brightness 13
building area method 198
building energy code 191, 204

**C**
CALiPER 129
candelas 11, 18
    per square meter 13, 14, 26
cavity ratios 53
CDM 93
ceiling cavity 53
ceramic metal halide 88, 93
choromaticity chart 42, 215
circulation areas 73
clerestory window 140, 141
closed loop 153, 154
CMH 93
code-language standards 191
coefficient of utilization 55
color quality scale 44, 85
color rendering index 43
color temperature 41, 42, 216
COMcheck 200
commissioning 156
connected load 145, 146, 148
Consortium for Energy Efficiency 102
continuous dimming 167, 168
contrast 15
    ratios 36
correlated color temperature (CCT) 42, 43, 216
CRI (color rendering index) 43, 44, 85
CRTs 20, 22

**D**
DALI 180, 181
daylight dimming 149, 150, 153, 158, 167
daylight harvesting 149, 153
daylight infiltration 38
daylighting 38, 60, 71, 135, 138, 139
daylight studies 60
degrees Kelvin 41, 42
digital controls 178

dimming ballast 96
direct glare 16
direct/indirect basket troffer 24
direct/indirect pendant 76
distributed controls 178
DMX 180
DMX512 180, 181
drive current 117, 118, 119
driver 111

E
effective illumination 136, 138
effective lumens 84, 105, 132
electrodeless 105
Energy Independence and Security Act of 2007 89
Energy Star 206, 207

F
floor cavity (HFC) 53
fluorescent 66, 85
fluorescent lamps 95, 100
footcandles 13, 14, 48, 68

G
GHG 204
glare 15
glare zone 18, 22
green building standards 212

H
halogen infrared 92
halogen lamps 91
HID 85, 95
high-efficiency troffer 24
high-intensity discharge (HID) 66, 93
high performance T8 (HPT8) 101
high pressure sodium 95
horizontal footcandles 30, 31
horizontal illuminance 70

I
IECC (International Energy Conservation Code) 192, 201
IES file 33, 58, 59
IES LM-79 128
IESNA LM-63 58
IESNA LM-79-08 123
IESNA LM-80-08 123
IGCC 213
illuminance 13, 48, 65
  determination system 69
  targets 65, 66, 69, 74
incandescent 66
  lamp 89
indirect lighting 23
instant start ballasts 101
International Energy Conservation Code 194
International Green Construction Code (IGCC) 213
inter-reflected light 48
inverse square law 51, 52

K
Kelvin 42, 216
kilowatt-hour 165

L
L70 118
lamp 83
law of reflection 33
LED 44, 85, 88, 106, 111
  arrays 125
  retrofit lamps 127
LEED 209, 210
life cycle assessment 86, 124
light emitting diode 111

light emitting diodes 106
light engines 106, 116, 119, 120, 125, 127
lighting design 3
lighting power allowance 146, 197
lighting power density 146, 197
light loss factors 55
light shelf 140, 141
Living Building Challenge 210, 211
load shedding 165, 166, 167
localized general lighting 76
low pressure sodium 95
LPDs 148, 197
lumen 48
 method 52
lumens 11
luminance 13, 24, 26, 32, 33
 calculations 32
 ratios 37, 39
luminous efficacy 84
 ratings 84
luminous intensity 18, 23, 32
luminous intensity chart 18, 24
lux 14, 48, 68

**M**

magnetic ballasts 100
magnetic induction 88, 102, 106
media façade 114
mercury 94, 96
mercury amalgam 105
metal halide 93
model codes 191
model energy codes, 219

**O**

observer characteristics 69, 70
occupancy sensor 158, 159, 160
open loop 153, 154

**P**

parabolic troffer 22, 24
Passive House 204
passive infrared 161
peak demand 165, 166, 167
phosphor 115
phosphorescent 96, 103, 114
photometry 120
Planckian locus 41, 215
point-to-point calculations 51
power factor 102
program start ballasts 101

**Q**

quality lighting 7

**R**

radiosity 59, 60, 62
ray tracing 62
reflectance values 53
reflected 13
reflected glare 16, 33
relative photometry 107, 108, 120, 122
REScheck 200
RGB 112, 217
room cavity 53

**S**

scheduling 163
self-ballasted 102
semiconductor 112
solid state lighting 111
source 83
space-by-space method 198
specular 16, 28, 35, 60, 62, 71
specularity 13, 60
SSL 106, 111
step-dimming 167, 168
sustainable design 7

**T**
target efficacy rating 84
target illuminance levels 73
task-ambient 74, 75, 77, 78
task area 73
task characteristics 69, 70
task importance 69
task lighting 74
thermal management 117, 119
three-wire system 178
time-of-day controls 163, 164
Title 24 201, 202

**U**
ultrasonic 161
uniformity targets 70
USGBC 209
U.S. Green Building Council 209

**V**
vacancy sensor 158, 159
VDTs 18, 20
veiling reflections 71
vertical footcandles 30
vertical illuminance targets 70
view window 141
visible spectrum 41
visual age 69, 70
visual comfort 29
visual comfort probability (VCP) 32
visual interest 29
volumetric lighting 23

**W**
white light 41, 42
wireless controls 183
work-plane 30, 73

**Z**
Zero Net Energy Buildings 204
zero-to-ten-volt (0-10v) 174
zonal cavity method 52, 53